THE
HORSE
CARE
MANUAL

THE
HORSE
CARE
MANUAL
CHRIS MAY

STANLEY
PAUL

A QUARTO BOOK

Copyright © 1987 Quarto Publishing plc

Stanley Paul & Co Ltd

An imprint of the Random Century Group
20 Vauxhall Bridge Road, London SW1V 2SA

Random Century Australia (Pty) Ltd
20 Alfred Street, Milsons Point, Sydney, NSW 2061

Random Century New Zealand Limited
191 Archers Road, PO Box 40-086, Glenfield, Auckland 10

Century Hutchinson South Africa (Pty) Ltd
PO Box 337, Bergvlei 2012, South Africa

First published 1987
Reprinted 1990

British Library Cataloguing in Publication Data

May, Chris
  The horse care manual: how to keep
  your horse healthy, fit and happy
  1. Horses
  1. Title
636.1'083   SF285.3

ISBN 0-09-172693-X

This book was designed and produced by
Quarto Publishing plc
The Old Brewery
6 Blundell Street
London N7 9BH

Senior Editor Helen Owen
Art Editor Ursula Dawson
Editor Mike Darton
Designer Richard Slater

Art Director Moira Clinch
Editorial Director Carolyn King

Typeset by Gatehouse Wood Ltd and Burbeck Associates Ltd
Manufactured in Hong Kong by Regent Publishing Services Ltd
Printed by Leefung-Asco Printers Ltd, Hong Kong

CONTENTS

SECTION 1

CARING FOR YOUR HORSE

GETTING TO KNOW YOUR HORSE

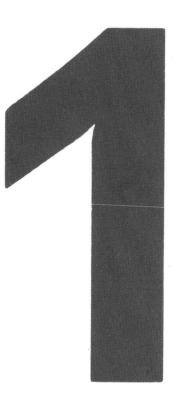

Horses and ponies are sophisticated and intelligent animals. Their anatomy has evolved to make them efficient grazers, and to provide them with speed and quick reactions, to escape from the danger which their acute senses are ideally designed to detect. Looking after horses involves not only supplying their physical needs, but also being aware of their mental well-being and appreciating their senses. Knowing what a horse can see and hear enables you to approach it without causing alarm. Getting to know your horse, its character and idiosyncrasies, should help handling, schooling and riding — and is one of the main attractions of owning a horse.

THE HORSE AND ITS ANCESTRY

The horse has gone through many evolutionary changes since it first appeared on earth, some 55 million years ago.

? WHAT IS THE HORSE'S POSITION IN THE ANIMAL KINGDOM — HOW ARE HORSES RELATED TO OTHER EQUINE SPECIES?

All domestic horses are members of one animal species, *Equus caballus*. The different varieties are each derived from the wild horses that once roamed central Asia, of which the sole surviving example today is the Mongolian Wild Horse, the so-called Przewalski (or Przhevalski's) horse. These were once thought to be extinct, but were rediscovered living in the Gobi desert in 1881, by a Russian explorer, Colonel Nikolai Przewalski (or Przhevalski). There is now a well-established zoo population of these animals, but there are thought to be no more living in the wild. The Przewalski is virtually a separate species of horse, having two fewer chromosomes than the domestic horse.

There are five other equine species (equids) surviving today: the common or plains zebra, the mountain zebra, Grévy's zebra, the African wild ass (the forbear of the donkey), and the Asiatic wild ass (the onager). Some equine species have become extinct only recently. The zebra-like quagga, which lived in South Africa, was wiped out by hunters as recently as during the 1870s.

The distribution of the horse's closest relatives today can be seen on the map. All six equine species can interbreed, but because the cells of each species have a different number of chromosomes, the offspring of such matings (for example, the mule, which is produced by crossing a male donkey with a mare) are infertile; crosses between Przewalski horses and domestic horses, however, are fertile.

Other presentday members of the order Perissodactyla (animals which have an odd number of toes) are, perhaps surprisingly, the rhinoceros and the tapir. Both of these, like the horse, have a false nostril in their nasal cavities, and walk on one toe (although the rhino has two and the tapir three relatively non-functional digits on either side).

Early Equus. After nearly 60 million years of evolution, this primitive horse developed to become the ancestor of all present day horses.

The Przewalski horse, or Mongolian Wild Horse. The ancestors of this animal were among the first horses to be domesticated, around 3,000 BC on the Russian steppes.

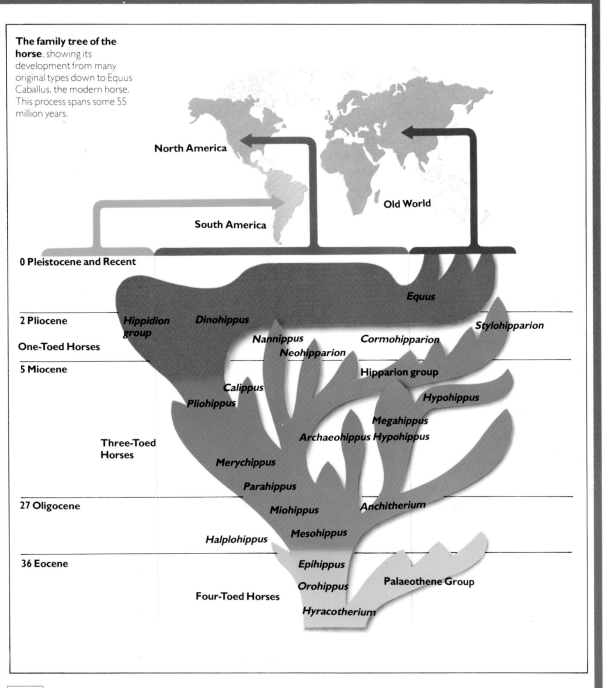

The family tree of the horse, showing its development from many original types down to Equus Caballus, the modern horse. This process spans some 55 million years.

North America

Old World

South America

0 Pleistocene and Recent

Equus

2 Pliocene

Hippidion group

Dinohippus

Stylohipparion

One-Toed Horses

Nannippus

Cormohipparion

Neohipparion

5 Miocene

Hipparion group

Calippus

Hypohippus

Pliohippus

Megahippus

Archaeohippus Hypohippus

Three-Toed Horses

Merychippus

Parahippus

27 Oligocene

Miohippus

Anchitherium

Mesohippus

Halplohippus

36 Eocene

Epihippus

Orohippus

Palaeothene Group

Four-Toed Horses

Hyracotherium

WHAT ARE THE ANCESTORS OF THE HORSE?

The evolution of the horse has been studied in considerable detail and, unlike the histories of many other types of animal, there is a fairly complete set of fossils of its ancestors' bones. The earliest mammals that can be considered to be ancestors of the horse are the Condylarthra. These primitive creatures first appeared about 75 million years ago, and are the forebears of all hoofed animals.

They were about the size of a fox — 36 centimetres (14 inches) at the shoulder — they had five toes on each limb, and lived on the succulent leaves of the small shrubs growing in the swamps covering the Northern Hemisphere at the time.

? HOW DID THE HORSE EVOLVE?

The process of evolution involves natural selection through the survival of individuals that are most suited to their environment. As the climate of the Northern Hemisphere changed, the swamps dried out and the ground became firmer. Walking on five toes, which had been an advantage in the swamps, became a disadvantage on the drier grassland (savannah) that developed on the plains. Over millions of years a gradual reduction in the number of digits occurred, until the first single-toed horse ancestor — Pleohippus — appeared 10 million years ago.

A steady increase in limb length and body size also took place. The small, relatively soft, short-crowned teeth of the earliest ancestors were ideally suited to the diet of swamp plants; they were totally inadequate to cope with the tough, indigestible grasses that grew on the savannah plains. So, over millions of years, a gradual lengthening of the teeth crowns and the development of their grinding surfaces occurred. The skull gradually enlarged to accommodate these developments to the teeth.

From North America, primitive equids spread south to South America and west to Asia (across the land-bridged Baring Straits). Large herds roamed central Asia from ten million years ago onwards. The first true horse — *Equus caballus* — appeared in America around one million years ago, and from there spread to Asia. In Asia, horses and the other equids were dispersed by the changes of vegetation which took place during the Ice Ages; horses, asses and zebras were pushed further south, into the Middle East and Africa. Over the centuries, horses developed into distinct strains of wild horse. The steppe horse thus occurred in Central Asia, and is the forefather of many of today's breeds (and also of the Przewalski horse). The plateau horse roamed Eastern Europe and the Ukraine, and from it developed the tarpan (another ancestor of many of today's warm-blooded breeds, and only itself becoming merged into ordinary horse breeds around a hundred years ago). In Northern Europe there lived two types of horse: the forest horse, forebear of the 'cold-blooded' breeds of heavy horse and the smaller, primitive, wild horse of the tundra, ancestor of many of today's pony breeds.

For some unknown reason (but possibly due to a fatal disease) all equids in America disappeared around 8,000 years ago.

Eohippus – the Dawn Horse. This four-toed animal, about the size of a hare, was one of the earliest equine ancestors. It lived in the swamps around 50 million years ago.

Thomas Huxley, the eminen biologist and supporter of Darwin's evolution theories *(below left)*. In 1876 he visited the USA, where he studied horse fossils collected by **O C Smith** *(right)*, Professor of Palaeontology at Yale University.

? WHAT EVIDENCE OF THEIR EVOLUTIONARY PAST IS PRESENT IN HORSES TODAY?

The 'chestnut' (a small piece of horn on the inside of horses' legs, just above the knees and on the lower part of the hock joints) is the remains of the first digit, lost during evolution. Likewise, the 'ergot' (the small horny growth on the skin of the back of the fetlock joints) is thought to be the remains of the horn of the second and fourth digits. Although these digits are otherwise absent externally, small remnants of the second and fourth metacarpal and metatarsal bones above are still present internally, in the form of the small, thin, splint bones which lie on the inner and outer edges of the upper part of the cannon bones (third metacarpal or metatarsal) on each limb.

WHEN WERE HORSES FIRST DOMESTICATED?

Prehistoric man hunted primitive wild horses for meat and for hides to use as clothing and shelter. It seems most likely that the first contact between man and horse occurred in Central Asia. Precisely when horses first became domesticated, however, is a matter of conjecture. The nomadic tribesmen of the steppes appear to have captured and reared foals, taking them with them on their travels, and eventually breeding from them. By doing this the nomads were provided with milk, meat, hides and a means of transport. But when this first happened will probably forever remain a mystery because these people had few possessions and left no records.

The first documented evidence of horses being ridden comes from China around 4000 BC, when mounted warriors were described, who came from beyond the northern border. These steppe horsemen became increasingly mobile, and used horses for migration in search of food, and for warfare. It was also in the steppes that the first spoked wheels and hubs were produced, innovations which allowed loads to be moved at relatively fast rates across the plains. The development of the chariot was then to revolutionize warfare and change the course of history.

In Western Europe, men were much slower to appreciate the advantages of using horses for anything other than as food. Stone-age men there hunted herds of horses by driving them over cliffs, and huge piles of bones resulting from this practice have been found in southern France. The practice of eating horses almost entirely died out through the spread of Christianity into Northern Europe in the 8th-10th centuries AD when, although eating horseflesh had for instance formed part of the pagan worship of the Norse god Odin, it was forbidden by the Pope.

EARLY MURALS

Early records of the horse's domestication. *(Right)* King Ashurbanipal of Assyria leading his troops. *(Above)* An Egyptian tax collector's wagon laden with corn.

? WHERE DO TODAY'S BREEDS OF HORSES COME FROM?

From the largest heavy horse to the smallest and lightest — from the Belgian (up to $19\frac{1}{2}$ hands high and weighing up to 1,450 kg or 3,200 lbs) to the tiny Falabella (only 65 centimetres or $25\frac{1}{2}$ inches high and so small that it cannot support even a child rider) — all have, in one way or another developed from the primitive Asiatic wild horse. These horses were similar in shape and size to the Przewalski horse, were dispersed by the Ice Ages, and some — native breeds of pony such as the Icelandic and Exmoor — have changed very little since the Iron Age.

Other types of horse were pushed further south and, under human influence, were domesticated and selectively bred for specific purposes. Larger animals were required in some civilizations for use in warfare, for pulling chariots, so their horses were bred for an increase in size. Toughness and stamina were necessary in the harsh conditions of the Middle East and North Africa, and selection for these attributes formed an integral part of the development there of the Barb and the Arab from native stock. The Bedouin were particularly involved in the advancement of the Arab breed. They took great care in retaining mares with the best powers of speed and endurance (rather than looks) for breeding, and in using the best-looking stallions to mate with them. Larger animals were again required for transport, and especially to carry knights in armour, during the Middle Ages. But although the collar was first used in the 11th century, and enabled horses to pull carts more effectively, the horse was not used in agriculture until the Industrial Revolution, when improvements in harness and mechanized farm machinery produced a demand for the power in traction that could at this time be provided only by heavy horses. During the last two hundred years, the heavy breeds of 'cold-blooded' horses — the Shire, Clydesdale and Suffolk Punch, among others — have thus been developed.

? WHAT IS THE EFFECT OF SELECTIVE BREEDING BY HUMANS?

The size, shape and performance of each of today's breeds of horse are entirely the result of selective breeding by humans. The widest influence on today's breeds, though, must surely be the Arab. Arabs have been used extensively to improve many native breeds of horse (such as the New Forest Pony and the Welsh Mountain Pony). The Arab has played a part too in the origins of the Thoroughbred, because the parentage of all Thoroughbreds in the General Stud Book can be traced to three Arabian horses (the Darley Arabian, the Byerley Turk, and the Godolphin Barb) which were imported into England in the late 17th century and crossed with the heavy types of horse then in use.

Another horse that has had a wide influence on today's European breeds is the Andalusian horse from Spain. This breed was itself derived from native Spanish stock crossed with Barbs introduced by Moorish invaders. When introduced into America by the Spanish invaders of Mexico under Cortez, some of the animals escaped into the wild, where they were able to live and breed. Eventually these resulted in the Mustang, from which, indirectly, the Quarter Horse has developed.

Cream

Fleabitten grey

Palomino

Strawberry roan

Grey

Dapple grey

Yellow dun

Blue roan

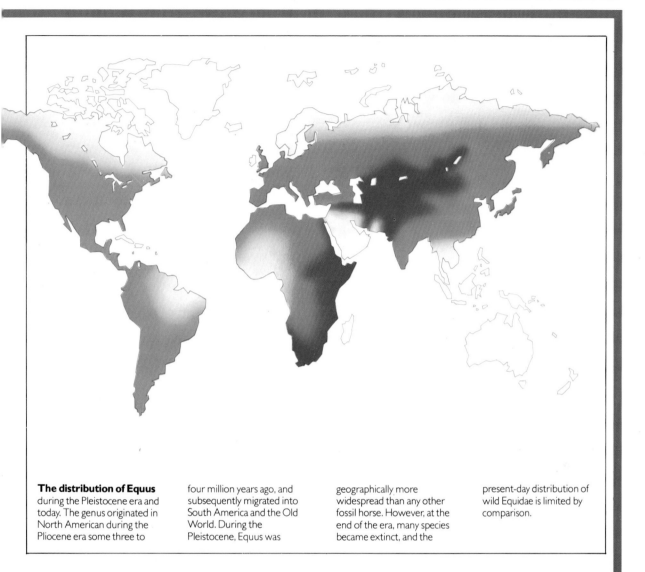

The distribution of Equus during the Pleistocene era and today. The genus originated in North American during the Pliocene era some three to four million years ago, and subsequently migrated into South America and the Old World. During the Pleistocene, Equus was geographically more widespread than any other fossil horse. However, at the end of the era, many species became extinct, and the present-day distribution of wild Equidae is limited by comparison.

Chestnut   Bay   Piebald   Odd coloured

Liver chestnut   Brown   Skewbald   Black

ANATOMY OF THE HORSE

Understanding a horse's anatomy and make-up is a great help in assessing its happiness and well-being.

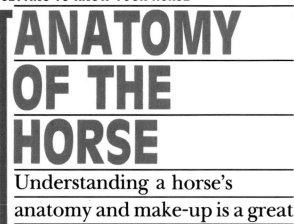

DOES A HORSE'S DIGESTIVE SYSTEM DIFFER FROM THAT OF OTHER ANIMALS?

The horse has to have a highly specialized digestive system in order to cope with its natural diet of grass — a relatively indigestible material.

Equine teeth differ from their human counterparts in that they are continually erupting throughout life to compensate for the equally continual wearing away of the surface by constant grinding. The incisor teeth have flat 'tables' which enable horses to graze very tight to the ground. A highly mobile tongue conveys food back to the cheek teeth, which have a very efficient grinding mechanism composed of interlocking ridges covered with enamel and cement.

The horse has a relatively small stomach for its size. Its capacity varies between eight and 15 litres (two and four gallons) in comparison with the cow's, whose four stomachs have a capacity of 137 to 274 litres (30 to 60 gallons). Although some digestion may take place in the stomach and small intestine, nearly all the digestion of grass occurs in the horse's very extensive large intestine, which by itself has a capacity of 80 to 130 litres (17.5 to 28.5 gallons). This organ contains vast numbers of bacteria which break down the cellulose and other tougher components of grass into volatile fatty acids, which can be absorbed. Some fermentation also occurs in the caecum, a dead-end compartment attached to the beginning of the large intestine. (Humans also have a very short caecum, which ends in the comparatively useless appendix.) Horses thus differ from the other main types of grass-eaters — the ruminants — in whom the breakdown of grass occurs almost entirely in the stomach(s).

Equine digestion functions by adding large volumes of fluid in the form of saliva and gastric juices — horses produce 10 to 12 litres (around three gallons) of saliva daily — to the ingested food at the beginning of the digestive tract. This fluid is reabsorbed, together with dissolved nutrients, at the end of the alimentary system. Sloppy droppings — which may appear as diarrhoea, normally resulting from excessive bowel movement — may thus be due to failure to reabsorb fluid through damage to (thickening of) the final part of the large intestine, the colon. Because of the anatomy of the palate, horses are unable to vomit. But, if the stomach becomes exceptionally distended, food may be passed up through the nostrils.

WHAT ARE THE 'POINTS' OF A HORSE?

Most occupations and hobbies have their own jargon: horse-keeping is no exception. The terms used to describe different parts of the equine anatomy, and the names of many of the diseases which affect them, have evolved during thousands of years' close contact between humans and horses — and for that reason the origin of many of these words is obscure. Nevertheless, although some of the disease-names — used in the past to intimidate the uninitiated — are disappearing in favour of more scientific terminology (generally providing a more accurate description of the cause of a problem), the names in common use for the external features of the horse's anatomy — the points of the horse (see diagram) — have remained unchanged for centuries.

The derivation of some terms, such as 'stifle' (the equine equivalent of the human knee-joint), is lost in antiquity, but some of the words come from mediaeval English (such as *fitlok* or *fetlak:* 'fetlock'), or originate from Old English (such as *hōh:* 'hock'); other terms have evolved through Old French — thus 'pastern' comes from *pasturon,* the tether used to tie up horses round this part of their legs when at pasture.

HOW IS A HORSE'S RESPIRATORY SYSTEM ADAPTED FOR SPEED?

Horses have a highly efficient mechanism for supplying oxygen to the muscles, thus facilitating both speed and endurance. It has been estimated that the total internal surface area of a

White face

Stripe

Blaze

Snip

Star

Identification In addition to the sex, height, age, breed, and coat colour, horses' markings are also recorded on veterinary certificates to help identify them. A drawing and a written description of the face and leg markings are used. Exact descriptions for limb markings, such as white to mid-cannon (*above*) are now used, rather than the previous vaguer terms such as 'socks' or 'stockings

The size and shape of the markings on the horse's face (*left*) are also a means of identification used when describing individual horses. Some of the more common ones are shown here. The size and position of a star, and whether a stripe is narrow or broad, should be stated. A star followed by a stripe is usually described as a disjointed stripe.

horse's lungs — the area available for oxygen uptake — is in the region of 2,500 square metres (nearly 27,000 square feet). This compares with 650 square metres (7,000 square feet) in a cow, and a mere 90 to 150 square metres (970 to 1,615 square feet) in humans!

The horse's head is made lighter by the presence of large air spaces within the skull bones (sinuses) which communicate with the airways. The horse also has two other unusual features in its upper airways, one in the nose and one in the throat region; their function is unknown, and because there are no soft tissues available from fossil records no one can say how or when they evolved. At the nasal entrance of the airways, the horse has an offshoot of each nostril — the false nostril. The horse's upper airway has what are

called the guttural pouches: two air sacs, one on either side of the pharynx (just below the ear).

Because of the length of the soft palate, horses are unable to breathe through their mouths, but the large nasal passages and windpipe are efficient in providing a good airflow to the lungs. The one weak point in this system is the narrowest point of the airway — the larynx. Narrowness between the lower jawbones, excessive head flexion (by the rider), or disease causing paralysis of the larynx ('roaring') may all restrict air intake and reduce the animal's performance. When galloping, horses are able only to breathe in phase with the stride, inspiration (inhaling) occurring when the forelegs are extended.

POINTS OF THE HORSE

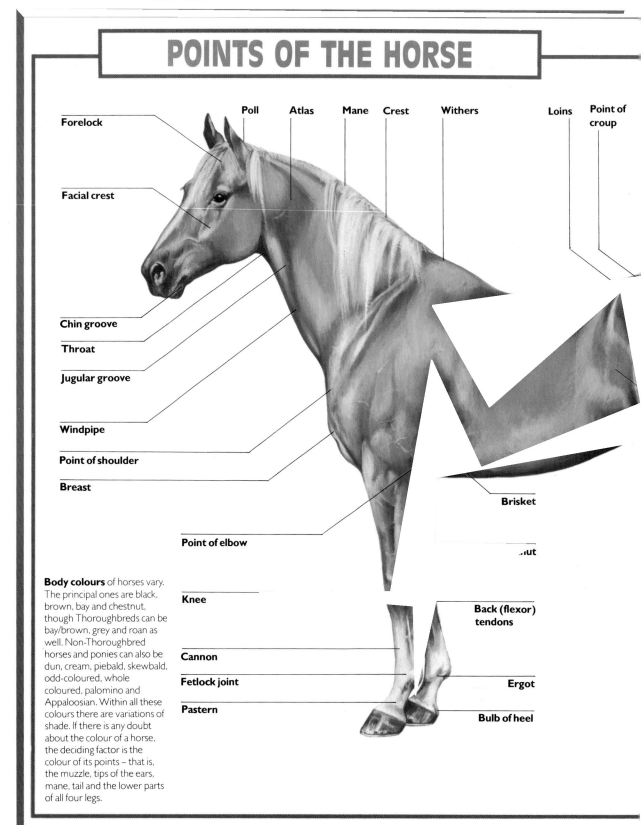

Forelock

Poll

Atlas

Mane

Crest

Withers

Loins

Point of croup

Facial crest

Chin groove

Throat

Jugular groove

Windpipe

Point of shoulder

Breast

Brisket

Point of elbow

...ut

Body colours of horses vary. The principal ones are black, brown, bay and chestnut, though Thoroughbreds can be bay/brown, grey and roan as well. Non-Thoroughbred horses and ponies can also be dun, cream, piebald, skewbald, odd-coloured, whole coloured, palomino and Appaloosian. Within all these colours there are variations of shade. If there is any doubt about the colour of a horse, the deciding factor is the colour of its points – that is, the muzzle, tips of the ears, mane, tail and the lower parts of all four legs.

Knee

Cannon

Fetlock joint

Pastern

Back (flexor) tendons

Ergot

Bulb of heel

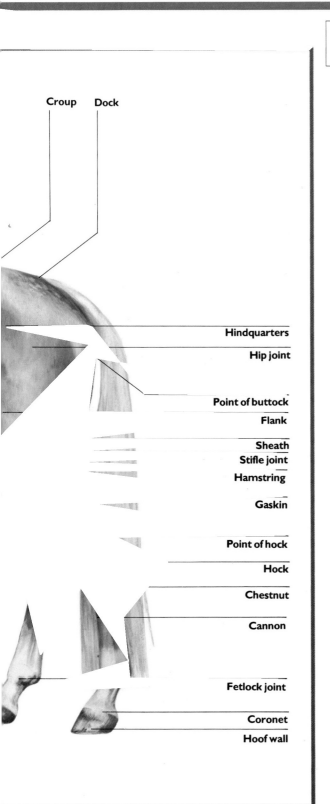

Croup Dock

Hindquarters

Hip joint

Point of buttock

Flank

Sheath

Stifle joint

Hamstring

Gaskin

Point of hock

Hock

Chestnut

Cannon

Fetlock joint

Coronet

Hoof wall

? HOW IS A HORSE'S BODY ADAPTED FOR SPEED AND ENDURANCE?

Two demands have influenced the evolution of the horse's locomotor system — to be able to run away quickly from danger, and to be able to cover large distances on relatively hard surfaces in search of food. Greater speed is possible because of the length of the.leg, and the single toe gives less strain. The concave sole and the frog protruding down from it help to provide a highly efficient grip on the ground. The horse's head and body are streamlined to reduce wind resistance, and the large muscles responsible for movement are situated high up on the limbs to reduce weight on the moving parts. The powerful effects of the muscles are transferred to their associated moving parts by a highly sophisticated system of tendons. Unlike other fast-moving animals (such as greyhounds or cheetahs), horses have a very restricted range of movements of the spine. In older horses, many of the spinal vertebrae actually fuse together, permitting little or no flexibility.

The principal problem associated with fast movement on a hard surface is concussion. A horse's anatomy contains many special features which enable it to overcome this problem. The main propulsive force comes from the hindquarter muscles. Most of the weight and the concussive effect, during motion, therefore fall on the forelimb. The horse's shoulderblades are attached to the chest by muscles which act as a supporting cradle, capable of absorbing much of the impact of the chest on the forelimbs during motion. The ligaments and tendons of the forelegs also act as a highly efficient shock-absorber. One has only to look at the slow-motion replay of show-jumpers competing in puissance events to marvel at the way in which the fetlock joints are able to sink to the ground, and spring back up again, to absorb the shock. They can do this because of a system of ligaments known as the 'suspensory apparatus' which involves the suspensory ligament, a ligament that runs directly behind the back of the cannon bone to insert on two small bones (sesamoids) at the back of the fetlock joint. From here, the ligament branches to various points on the back and front of the pastern bones, and it is able to support the fetlock joint when it sinks on each stride.

In the hind limb, the hock and the stifle (the equine equivalent of the knee) joints move together in unison because of the action of the muscles and ligaments linking the bones on each side of these joints. This combination too acts as a shock-absorbing system. The anatomy of the hind limb also provides a horse with another feature that enables it to make a rapid escape from danger — its ability to 'sleep', or at least to rest, while standing. Unlike humans and most domestic animals, who have only one ligament running from the kneecap (patella) to the bone below the knee joint, horses have three patella ligaments in their stifle joints. By contracting its thigh muscles, a horse is able to 'lock' its stifle joint, relax the muscles, and rest while standing.

Investigation of equine muscle tissues has shown that as in most animals there are different types of muscle fibres present in them. Some of these ('fast-twitch' fibres) are responsible for muscle work in the absence of oxygen, as is required for short spells of movement at high speeds (sprinting and escape from danger). Other types of fibre ('slow-twitch') are responsible for slow, sustained, repeated muscular contraction which requires oxygen (endurance, stamina, staying power). The relative proportion of one type of fibre to another in any individual horse's muscles is determined by heredity. Thus in racehorses, sprinters have relatively more fast-twitch fibres, and stayers have a higher percentage of slow-twitch fibres. When muscles contract, particularly during exertion in the absence of oxygen, lactic acid is produced. Accumulation of

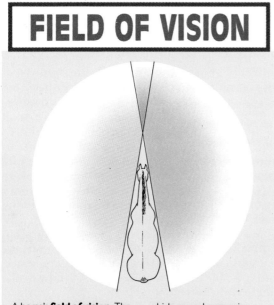

FIELD OF VISION

A horse's **field of vision**. The shaded area *(above)* represents the area of binocular vision, which, as can be seen, is extremely limited.

this chemical is largely responsible for muscle fatigue. Training a horse, and getting it fit, mainly involves developing the efficiency of the different types of muscle fibre, and their ability to cope with a build-up of lactic acid. Specific exercise schedules can be helpful in developing the different types of muscle fibres required for different forms of work or competition.

 WHAT ARE THE NATURAL PACES OF A HORSE?

Horses have four distinct forms of movement — although some intermediate gaits are also possible. These are the walk, the trot, the pace and the gallop. At the walk, the four legs are placed to the ground in regular succession — the sequence of footfalls is left fore, right hind, right fore, left hind. This is a four-beat gait; when listening to a horse walking on a hard surface, four distinct sounds should be heard.

The trot, however, is a two-beat gait; the diagonal legs are moved synchronously. The footfalls in sequence are left fore with right hind and right fore with left hind.

At the canter, one foreleg leads while the other foreleg and its diagonal hindleg move together, and the other hindleg moves independently. This is thus a three-beat gait, with a footfall sequence (with a left fore lead) of right hind, then left hind with right fore, followed by left fore. There is a period of suspension after the leading foreleg leaves the ground.

In the gallop, the stride lengthens and the period of suspension is also increased. At the same time, the legs that were working in a diagonal at the canter are unable to do so at this faster pace. They become separated, so that the hind leg hits the ground slightly before the diagonal foreleg. Hence the footfall sequence (with a left fore lead) is right hind, followed by left hind slightly before right fore, followed by left fore. The gallop is therefore a four-beat gait. In the canter and gallop, either foreleg may lead while going in a straight line, but it is usually the inner leg that leads on a bend. The leading leg has to take more weight and do more work than its opposite number. Tired horses often change legs in mid-gallop for this reason.

WHAT CAN HORSES SEE?

Horses have large, prominent eyes, placed well to the side of the skull. This gives them very good all-round vision without having to move their heads. Such a configuration is common in herbivores, and enables them to detect the movement of predators while they are grazing. The opposite is true of carnivores, which have eyes placed close together at the front of the skull to give the binocular vision essential for them to judge the distance to their prey. Horses have a very limited field of binocular vision, and thus have difficulty in judging distance. However, they can detect moving objects a long way away with considerable ease. Horses have two blind spots. The first is directly behind them, and the second is directly in front of the end of the nose. They are unable to see the end of the muzzle, and the tactile hairs around it are important for locating food and other objects.

An appreciation of the horse's eyesight is useful to owners: approaching from behind or from directly in front should be avoided if you do not wish to startle your horse. A horse's eyes are prominent, and are therefore prone to injury, and it is a natural reaction for it to try to protect them. It may well be that shying away from an object (a drain in the road) or a sudden movement (a bird in the hedge) is a reflex protective mechanism derived from the horse's inability to see in front of its nose, or to accurately judge its distance from the object.

To get a better look at something unfamiliar, a horse usually moves its head so that it can see it with both eyes simultaneously.

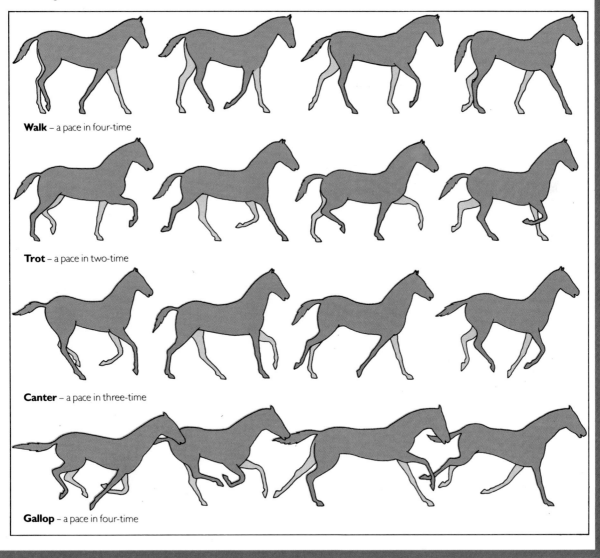

Walk – a pace in four-time

Trot – a pace in two-time

Canter – a pace in three-time

Gallop – a pace in four-time

BEHAVIOUR AND SENSES

An appreciation of horses' senses and how they live in the wild can give an insight into the way they behave

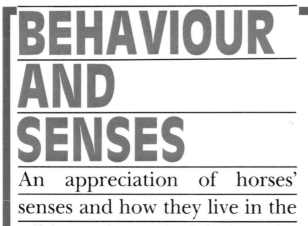

? WHAT ARE THE NORMAL BEHAVIOURAL PATTERNS OF HORSES, AND HOW ARE THEY AFFECTED BY DOMESTICATION?

The behaviour patterns of wild horses and horses that have escaped into the wild have been studied extensively. Horses are essentially herd animals. In nature, a herd is small, comprising of three to eight mares, their foals, and the herd stallion. There is always a hierarchy among the mares, and the 'lead' mare is usually the one who dictates when the herd grazes or rests, rather than the stallion, whose role is mainly protective. In the wild, there are also bachelor herds of stallions, in which no particular animal appears to be dominant. Colts and fillies leave the herd at between two and three years old, fillies frequently being removed by bachelor

COMPANIONSHIP IN THE FIELD

Lipizzaner mares and foals. This breed *(above)* is descended from Andalusian horses imported from Spain, and takes its name from the stud founded at Lipizza in about 1850 by an Austrian archduke. The foals are born with dark coats which do not turn grey until they are at least seven years old. Although adult horses can also be chestnut or black, only grey stallions are used by the Spanish Riding School in

Vienna to perform the intricate movements of *haute école* equitation.

Mustang foals. These native feral horses of North America *(right)* are also descended from Andalusian stock. There were no horses on the American continent until the Spanish invasion of Mexico in the sixteenth century. Horses escaping into the wild from this expedition became the forefathers of the Mustang.

stallions to begin a herd of their own. In wild horses, there is very little importance attached to territory by stallions, because herds migrate over large distances. This is in direct contrast to asses, of which the males defend their territory from other males and mate only with females entering it during the breeding season. It is hard for humans to appreciate the importance of this herding instinct in horses. As well as providing protection, it also has other benefits, such as mutual grooming, and possibly protection from flies when horses gather together.

HOW IMPORTANT IS COMPANIONSHIP TO HORSES?

This is hard to assess without venturing into anthropomorphism (giving animals human characteristics). However, horses always appear to be 'happier' when in the company of other horses, and some horses do appear to suffer without company. Perhaps they miss the contact and benefits of such behaviour as mutual grooming. Such animals (particularly youngsters) can pine and refuse to eat. They may be helped by companionship with an animal other than a horse if none is available. Sheep, goats, cats, rabbits and hens have all successfully filled this role.

WHAT FRIGHTENS HORSES, AND HOW DO THEY RESPOND?

A horse's reaction to danger — real or imagined — is to run away. Because of their blind spots and the difficulty in judging distance with one eye, horses frequently make as if to run away (shy) when faced with a sudden movement or unfamiliar object. For the same reason, it is important to talk to a horse when moving around it so that it can be sure of where you are. Likewise, running a hand from a part of the horse's body where it can be seen, along to a part where it cannot, is better than touching the spot suddenly and startling the animal. Whereas rearing can be a sign of fright, biting and striking with a foreleg are signs of aggression rather than fear.

WHAT IS MEANT BY 'TEMPERAMENT', AND IS THIS AFFECTED BY BREEDING?

A horse's temperament is its general demeanour and the manner in which it responds to its owner or rider. That response can include both willingness to be handled, and to submit to control when ridden. Breeding has a large influence on temperament. Some breeds or types of horse (such as cobs or heavy horses which are sometimes called 'cold-blooded') are naturally phlegmatic, and usually respond favourably to human handling. Others (such as Thoroughbreds) are much more likely to be 'hot', bad-tempered, and to require skilled handling to control them. Temperament can be made worse by injudicious handling or through boredom and insufficient exercise. Many ponies are made 'nappy' by spoiling them.

Evaluating temperament is vital when buying a horse, and time spent observing how an intended purchase behaves when being caught, tacked up or ridden is never wasted. An ideal temperament depends on what the horse is to be used for, and the skill and confidence of the rider. A horse that is difficult to catch, kicks or bites in the stable, will not load, and puts its ears back and rears when ridden, is hard work and no enjoyment. Conversely, one that is easy to catch, quiet and sensible in the stable, yet alert and interested when ridden, is a pleasure to own.

CAN HORSES SEE COLOURS?

Tests have shown that, unlike some animals (such as cattle and dogs which can distinguish only different shades of grey), horses can distinguish colours. They are readily able to pick out green and yellow, but are less proficient at distinguishing red and blue from grey of the same intensity.

WHAT CAN HORSES HEAR?

Horses have a highly developed sense of hearing. They also have extremely mobile ears which can move independently, rotating through an angle of 360 degrees; without moving, a horse can thus pick up sounds from all directions.

Horses can also hear sounds of a pitch too high to be audible to the human ear, and are able to distinguish specific words rather than the tone in which they are spoken; they are additionally much better able to judge the location of the source of a sound than humans. For this reason it is important to talk to horses when moving around the stable or approaching them: they are using mostly sound, rather than eyesight, to detect your whereabouts.

WHAT OTHER SENSES ARE IMPORTANT TO HORSES?

Horses have very well-developed senses of both taste and smell. They are capable of differentiating between different grasses and weeds in a paddock. They can also distinguish preferred foods from disliked or doctored feeds in a manger, with great accuracy. Horses seldom eat poisonous plants for this reason, unless they are cut and wilted. Stallions can scent a mare that is in season but out of sight several fields away, and the behaviour they display (curling up the upper lip) is also used by horses of both sexes on other occasions, presumably to enhance the sense of smell. Smelling a strange horse or its excreta (often accompanied by snorting) is important in social contact between horses, as is recognition of a foal by its dam. Smell is also used to investigate unfamiliar objects, again accompanied by snorting, which may enhance this process.

The sense of touch is very important to horses. The end of the muzzle is especially sensitive, and a horse uses this to investigate strange objects. The long hairs there have an important tactile function, which may compensate to a great extent for the animal's inability to see in this area.

WHAT VICES CAN HORSES DEVELOP, AND WHY DO THEY DO THIS?

Nearly all vices develop in stabled horses as a result of boredom. A horse's natural habitat is the wide open spaces, and no matter how large the stable, horses do not like being confined for long periods without exercise. Pawing the floor or door, or kicking the walls are common problems. Boredom may also be responsible for wood- and rug–chewing, and also for 'weaving' (rocking from one leg to another accompanied by side-to-side head movements — usually over the stable door). Crib-biting (catching hold of an object, perhaps a door or manger, with the incisor teeth, and sucking air into the stomach) or wind-sucking (arching the neck and inhaling air into the stomach without actually biting on anything) can both cause digestive upsets and make the horse 'poor'. Putting horses in larger yards, providing more roughage (hay) in the diet, or a companion animal, and increasing exercise may all help to overcome boredom. Specific preventive measures can also be taken. These may include removing objects used for crib-biting, painting wood with obnoxious pastes to prevent chewing, or putting grilles on upper stable-door openings to prevent

'weaving'. The development of some vices can be associated with psychological upsets, such as traumatic weaning. Continual box-walking is often found in nervous horses, and may be improved by companionship. Turning a horse out to grass usually stops boredom and cures many vices. However, bad habits such as wood-chewing and crib-biting can persist at grass, and may be copied by other horses!

DO HORSES POSSESS ANY SPECIAL SENSES?

It seems that horses are able to detect vibration in the ground. Owners will be aware of the way in which horses stand with all four feet square apparently rooted to the ground, and are somehow able to detect the approach of another animal before it can be seen or heard. Horses are said even to be able to detect fear in a rider! (In that case, it is more likely that their acute sense of smell has detected perspiration.)

HOW INTELLIGENT ARE HORSES?

Assessing intelligence in animals is difficult. Whereas humans work through the possible solutions to a problem in their heads and select the most likely answer, animals tend to try to solve problems by practical experimentation. Thus the ability to solve problems, and to remember and learn from the results, must be assessed in animals rather than their powers of reasoning. Horses are quite good at solving problems like opening gates. They also have a good memory, which can help them find their way home. Likewise, horses can learn quite readily, and can be trained to achieve a desired response to commands or equitation 'aids'.

When assessing intelligence, a horse's highly developed senses must always be taken into account. A good example of this was a horse called Clever Hans, which was able to count and do sums by stamping its forefoot. However, when the person with the horse did not know the answer to the question, Clever Hans was perplexed. What was eventually discovered was that when the horse stamped out the correct answer it was picking up some unconscious movement or vibration made by the questioner. Such a reaction did not happen when the person did not know the answer. This illustrates the acuteness of horse's senses, and the hazards of trying to assess their intelligence!

STABLE VICES

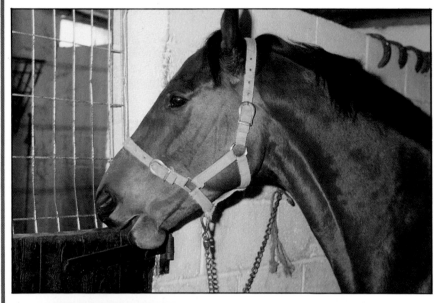

'Crib biting'. Horses suffering from this vice *(left)* catch hold of an object such as the stable door, manger, or a fence post with their teeth, arch their neck and inhale air into the stomach. Boredom is the most frequent cause, but the habit can be learned from other horses. Some horses, known as 'wind-suckers', even learn to arch their necks and inhale air into the stomach without first having to catch hold of an object with their teeth. Horses with these complaints tend to be nervous and are always hard to keep in good body condition, so it is best to avoid buying them, wherever possible.

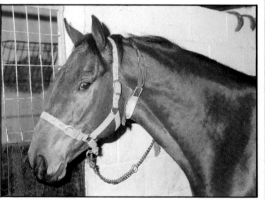

Excess wear of a horse's front (incisor) teeth resulting from 'crib biting' *(left)*. This may make it more difficult to tell the animal's age accurately. Inhaling air into the stomach interferes with normal digestion, which is why crib-biters and wind-suckers tend to lose condition.

A crib-biting collar. A leather strap attached to a specially shaped metal plate is put round the horse's neck to prevent 'crib biting' *(above)*. It puts pressure on the muscles which the animal uses to arch its neck when attempting to suck air, and will stop some horses. Removing all objects which the horse can catch hold of with its teeth is an additional preventive measure.

A HOME FOR YOUR HORSE

The first prerequisite for looking after a horse or pony is somewhere safe and suitable to keep it. Grazing and some form of shelter are required, but how much time the animal spends at grass or in the stable depends on the type of horse, the amount of work it is doing, the time of year, and how much time the owner has to look after it. Although certain factors — such as the choice of bedding material — are optional, there are minimum requirements in stables and paddocks that an owner must be aware of to ensure both health and safety. By making sure these are met, many injuries and health problems can be avoided.

THE OPTIONS

Every horse owner must consider carefully the conditions in which he will keep his horse. The facilities available will have a bearing on the type of horse chosen.

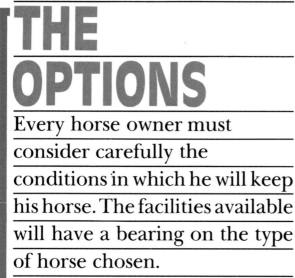

? STABLE OR PADDOCK?

It is far simpler, and more economical in both labour and cost, to keep a horse at grass for as much of the time as possible. A suitable paddock must have adequate grazing, water and shelter. The owner need then only visit the animal once a day to check the water, provide the feed, if necessary bed down the shelter, and generally ensure that the horse is healthy. Whether it is feasible to keep a horse or pony permanently at grass depends upon its breed and the weather conditions. Native ponies are much hardier than horses and in our climate can usually live outside all the year round. Thin-skinned animals, especially Arabs, Thoroughbreds and crosses from them, are not very hardy and must be stabled at night during the winter.

Keeping a horse in a stable requires a considerable commitment in both finances and time. Mucking out, feeding and grooming involve at least an hour's work a day, every day, and the need for regular exercise takes up a further 20 minutes to two hours. It may be necessary to keep a horse stabled or at livery if the owner has no access to grazing. Horses should not be stabled indefinitely; a spell at grass improves an animal's coat, hoofs and temperament. Stabling is often necessary in winter, and is essential for horses if they are to be got fit for competition or are clipped for regular work. Rations can then be carefully controlled. A horse that is exercised hard needs concentrates, not a stomach full of grass.

Suitable stabling and grazing. Most horses will need both stabling and grazing at some stage during the year. Points to look for in a stable are an adequate size, good ventilation (without draughts), warmth, and ease of cleaning. A stable with concrete walls *(left)* is ideal and easy to keep clean. Wood shavings provide a warm and comfortable bed. In a paddock, a good, even covering of meadow grass is desirable. Some shelter, such as that provided by large trees *(above)*, and fencing that is secure and safe for horses, is also essential.

WHAT PROBLEMS OCCUR WHEN A HORSE IS AT GRASS?

A horse at grass should be inspected daily for signs of illness or injury (see Chapters 4, 7 and 8). It is also necessary to keep a regular check on its body condition. The quality and quantity of grass can vary greatly throughout the year, and extra feed may be required to prevent a horse from losing condition. In colder weather a shelter or a New Zealand rug may be necessary. Wet weather can produce skin problems such as mud fever and rain scald. The risk of injury from kicks is reduced by removing shoes, especially the hind shoes. However, a horse's feet tend to break away without shoes at grass, and may benefit from shoes or 'tips' at least on the front feet. Horses at grass also need their feet trimmed from time to time and must be wormed regularly.

WHAT PROBLEMS OCCUR WHEN A HORSE IS STABLED?

Digestive problems are more common in stabled horses than in those at grass. Colic due to impaction — the internal solidifying of faeces due to slow bowel movement — is a particular danger, and bran mashes and other forms of laxative are commonly given to prevent it. Many horses are allergic to moulds in hay or straw. In the confined space of a stable, especially if ventilation is inadequate, such an allergy all too often produces a lung reaction and coughing in symptoms known as COPD, or broken wind. This disappears if the horse is turned out to grass. Circulation problems such as swollen legs and, more rarely, lymphangitis (inflammation of lymph vessels) may also be associated with stabling.

Perhaps the most common problem is boredom. This can be responsible for wood-chewing, crib-biting, weaving, box-walking and other stable vices. Less often, a nervous horse confined on its own, may pine and refuse to eat properly. Introducing a sheep, goat or other stable companion if possible may alleviate this problem.

MY HORSE IS STABLED ONLY PART OF THE TIME: IS IT BETTER TO KEEP IT IN AT NIGHT OR BY DAY?

A half-and-half system is an ideal compromise. The horse gets exercise at grass, but is easier to keep clean when stabled part of the time. Stabling also prevents having to catch the horse before riding. A combination of stabling and paddock involves less work than keeping a horse stabled continually. The horse need only be visited night and morning to be fed, watered, checked, and to get it in or out.

The best solution is to keep the animal in at night in winter to keep it warm and to give it extra feed, and to let it out to graze and exercise in the daytime. In the summer it is better to stable the horse during the day, and to turn it out at night. The reason for this is that in hot weather horses seek shade and shelter from pestering flies, and do not graze. When grass is lush it may in any case be necessary to restrict access to grazing, especially if the horse suffers from chronic laminitis (inflammation of tissue within the hoof). Stabling during the daytime may correct this condition. Stabling during late afternoon and evening is certainly essential to prevent sweet itch in individuals that are allergic to the midge bites that cause it.

INDOOR STABLING

Stabling for horses may be in the form of self-contained loose boxes or stalls housed in a barn. Either way, the accommodation must conform to certain standards.

Loose boxes with metal strips on the top of the stable door to stop horses chewing the wood *(above)*.

? WHAT IS THE IDEAL STABLE LIKE?

Many factors must be taken into account when choosing a stable. In particular, its distance from the horse's owner's home, and from the horse's grazing and exercise areas must be considered. Also important are a lack of traffic locally; the ease of access for delivering bedding and fodder and removing muck; storage facilities for food, equipment and tack; drainage; and the availability of water and electricity. It is obviously cheaper to use or convert an existing building wherever possible. Stables do not have to be tailor-made for horses. Many old buildings are satisfactory as long as they are of sufficient size and made of suitably strong material. However, they must be warm, draught-free, light, well ventilated and well drained, with no dangerous projections such as beams protruding.

? WHICH CONSTRUCTION MATERIALS ARE SUITABLE?

New stables should be constructed facing south for warmth, and in a sheltered position. Security must also be taken into account. It is best for a stable to be within earshot of a house — preferably your own. Remember that in most countries planning permission is needed for new stables; in Britain four copies of the plans are required, giving full details of the stables, relation to other buildings and local roads.

Older stables and buildings were commonly made out of brick — which is not as warm as most modern materials, so extra internal insulation may be needed in such constructions. New stables

built of concrete blocks should have cavity walls, preferably filled with an insulating material such as polyurethane foam. Because of the high cost of building, sectional wooden and prefabricated concrete stables have become popular; these are assembled on a concrete base and are perfectly adequate. An internal timber lining of exterior-grade plywood 18 mm (⅔ inch) thick or oil-tempered hardboard may provide protection from kicking and chewing, as well as giving extra insulation. Internal divisions or partitions should be at least two metres (seven feet) high.

Many materials are suitable for roof construction, but roof insulation is essential for warmth. This can be provided by an underlining attached to the exterior roof cladding, or by fitting a flat false ceiling. Ideally, the ceiling should be 3 metres (10 feet) high, and certainly not less than 2.5 m (8 ft).

Stalls Horses can be kept indoors in stalls, in which they are separated from one another by side partitions.

? WHAT SIZE OF STABLE DOES MY ANIMAL NEED?

The size required is directly related to the size of the horse or pony. Some recommendations are given in the accompanying table. In addition to ventilation problems, horses can get cast — that is, stuck against the wall and unable to get up — or injure themselves in stables that are too small. It is better to err on the side of providing more space than really necessary rather than too little, but stables should not be so vast that they are cold and draughty. A good size for a stable is 4.5×3.5 m (14×12 ft), with a height of 3 m (10 ft). Ponies require less space than this, and foaling boxes should be larger.

? WHAT KIND OF FLOORING IS BEST FOR STABLES?

Drainage is essential to allow urine to run away from underneath bedding so that it does not form pools and saturate it. Many old-fashioned stables had cobbled floors which were hard to clean thoroughly but did allow urine to drain away between the stones. Modern concrete floors are perfectly adequate as long as they have a slight slope and a shallow gutter to collect fluid, leading to a drain in the corner of the box or, better still, through the wall to an outside drain. If shavings, sawdust or peat bedding is used, drains must be covered to prevent them from becoming blocked. A new concrete floor should have a roughened surface to give a horse with shoes on some grip — no floor should ever be smooth. In areas which have a chalky soil, the chalk surface alone may be suitable for flooring, for it drains well.

STABLE FITTINGS

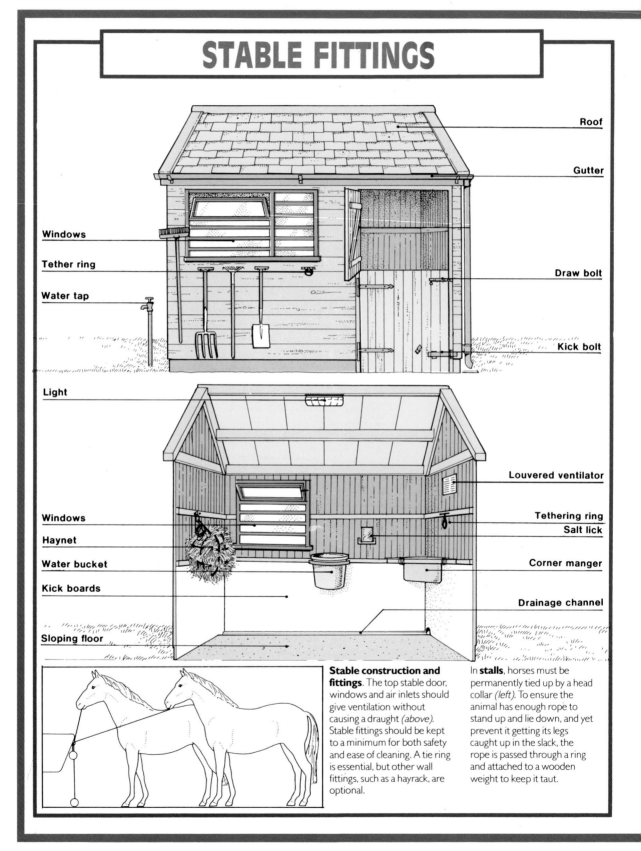

Roof

Gutter

Draw bolt

Kick bolt

Windows

Tether ring

Water tap

Light

Louvered ventilator

Tethering ring

Salt lick

Corner manger

Drainage channel

Windows

Haynet

Water bucket

Kick boards

Sloping floor

Stable construction and fittings. The top stable door, windows and air inlets should give ventilation without causing a draught *(above)*. Stable fittings should be kept to a minimum for both safety and ease of cleaning. A tie ring is essential, but other wall fittings, such as a hayrack, are optional.

In **stalls**, horses must be permanently tied up by a head collar *(left)*. To ensure the animal has enough rope to stand up and lie down, and yet prevent it getting its legs caught up in the slack, the rope is passed through a ring and attached to a wooden weight to keep it taut.

WHAT BASIC STABLE FITTINGS ARE NECESSARY?

A stable door must be at least 1.5 metres (4 ft) wide so that there is plenty of room for the horse, with its tack, to go through. It is all too common for a horse to damage the point of one hip by knocking it on a narrow doorway. Ideally, the door should be at least 2.5 metres (8 ft) high to allow plenty of head room; it should open outwards and be divided in half so that the top half can be fastened open for ventilation. The bottom half should be high enough to prevent a horse from getting over it, but low enough to be able to put its head over and look out. If necessary, however, a wire mesh grille can be fitted to stop the horse putting its head out. Fitting a grille with a central area only large enough for the horse's head prevents a horse from weaving. A roof overhang above the door is important to prevent rain from blowing in. Doors must be fitted with proper stable catches at the top. A second bolt or 'kick catch' at the bottom is an important precaution against escape artists, and a galvanized metal strip along the top of the bottom door prevents wood-chewing.

Windows should, ideally, face south for warmth. They should also be situated on the same wall as the door, and above head-level to prevent draughts. For the same reason, they should be hinged at the bottom and open inwards. They must always have a wire grille or bars fitted inside to prevent injury. Electric light sockets should also have a protective metal grille to prevent hay or bedding from coming into contact with a hot bulb. All cables must be out of reach of the horse, as must the light switch, which should be waterproof.

WHAT ADDITIONAL FITTINGS ARE NEEDED?

Whether other stable fittings are necessary is a matter of personal choice. The fewer fittings there are, the easier it is to clean the stable thoroughly. A well-secured tie ring attached to the wall at a height of 1.6 metres (5 feet) is useful. For safety, a piece of string should be attached to the ring, and the horse tethered to the string. If frightened, the horse will then break the string rather than breaking the head collar or injuring itself.

The tie ring can also be used for the haynet, which should be hung at horse's eye-level. Hay can also be fed on the ground. Wall hay racks above head-level are not advisable as animals tend to get

hay seeds in their eyes. Wall-mounted mangers require more cleaning and are another possible source of injury. Removable feeding bowls on the floor are perfectly adequate, and allow the horse to eat at a natural angle. Automatic water drinkers save work but give no idea of how much a horse is drinking: a plastic or rubber bucket with a capacity of at least 9 litres (2 gallons) standing on the floor is quite sufficient. Regular inspection of all fixtures, fittings and bucket handles is essential to make sure there are no sharp projections which could cause injury.

HOW CAN I BE SURE VENTILATION IS ADEQUATE?

The importance of adequate ventilation cannot be over-emphasized. A good supply of clean, fresh air is vital, not only to allow a horse to breathe and to remove the ammonia smell from decomposing urine and faeces, but to keep down the level of allergy-inducing fungal spores in the dust from hay and straw. Some idea of the quality of ventilation can be gauged from the smell of the stable — which should be fresh and pleasant. As a rule, it is better to supply plenty of fresh air and to keep the horse warm with rugs and blankets. However, ventilation must be controlled and directed, as horses may catch pneumonia if left in a draughty stable after exercise. A large part of the ventilation requirement can be achieved by keeping the top half of the stable door open. This should be possible throughout the year, unless the weather is exceptionally cold. Windows also provide ventilation and should ideally be on the same wall as the door, in order to avoid draughts. Technical information on the correct size and positioning of air vents, louvred inlets, roof outlets etc. is available from equestrian associations.

WHAT OTHER TYPES OF STABLING ARE AVAILABLE?

As well as individual stables, barns or buildings with many loose boxes are often used for stabling. These boxes generally each have a sliding door that opens into a central passage. They tend to be warm, and all stable routine can be carried out in relative comfort. However, the ventilation is often inadequate and so respiratory diseases tend to spread easily.

Less commonly, horses are kept in stalls, tied by a rope from their head-collars. Stall floors are sloped to the rear to aid drainage.

? WHAT BEDDING MATERIALS ARE AVAILABLE?

Bedding is necessary to allow a stabled horse or pony to lie down and rest in comfort. It also prevents its feet becoming jarred on a hard floor and encourages it to urinate (stale). A wide variety of materials can be used, either fresh (changed at least once daily) or as a deep litter that may not need changing for months at a time. Straw is the most commonly used material, and is probably the best as long as it is of good quality and not mouldy. Traditionally, wheat straw is preferred, mainly because barley awns can irritate a horse's skin. With modern harvesting methods this is less of a problem, although horses do tend to eat barley straw. Oat straw is not suitable as a bedding material because horses like to eat large amounts of it. Quite a number of horses are allergic to the moulds present, to a greater or lesser extent, in straw and hay, and they therefore require a different type of bedding.

Shredded paper is absorbent and makes a good clean bed, but it is expensive. Shavings are better than sawdust, which tends both to pool urine and to heat up when soiled. Peat moss is also popular, although it too is expensive and may be dusty when first used, but it has a low fire risk. Leaves and bracken are sometimes used for bedding because they are cheap. Leaves, however, quickly become saturated, and bracken can cause digestive problems if eaten even in moderate amounts, so neither of these is suitable.

Mucking out
Keeping your horse's bed clean and fresh is a chore you must never neglect *(above)*. A major clean-out should be done regularly, as well as a daily removal of soiled bedding and droppings.

? MUST I ALWAYS USE FRESH BEDDING?

No, deep litter means much less daily work. Straw, peat, shavings, sawdust and shredded paper can all be used in this way. After removing fresh droppings and damp shavings or peat in a skep (a wicker or plastic bucket), the surface of the bed is raked and fresh material added when necessary. When the bed reaches about one-third of a metre (one foot) in height it must be cleaned out. Shavings should be cleared every six months. Deep litter beds are warm but their decomposition produces many fungal spores. Any potential advantage in using a deep litter of shavings and shredded paper is thus completely wasted on horses suffering from COPD, and these materials must be used fresh for such animals. Foot problems — particularly thrush — are more common in horses bedded on deep litter, and extra care must be taken to pick out the feet regularly.

Wheat or barley straw
provide a good warm bed for horses. If banked against the walls, straw will help to stop draughts and prevent the horse getting 'cast'.

? WHAT CAN BE DONE TO PREVENT HORSES FROM EATING THEIR BEDDING?

If straw-eating is a problem, the bedding can be treated with a very weak solution of disinfectant sprayed on with a watering-can. This is usually enough to stop the habit. Otherwise it may be necessary to switch to an alternative type of bedding, such as shavings, peat or shredded paper.

THE PADDOCK

Finding suitable grazing for a horse that is to be kept out of doors can be hard. There are many considerations to bear in mind.

WHAT SHOULD I BE AWARE OF WHEN LOOKING FOR GRAZING FOR MY HORSE?

A horse or pony's paddock should be safe and secure, with adequate grazing, a fresh water supply and some shelter. Permanent pasture is the type of land most suitable for horses: they like old-established grasses and deep-rooted weeds rather than newly-sown lush grass which can cause digestive and other problems such as laminitis. Permanent pasture also has a good turf and tends to become sodden and trampled ('poached') less easily than land that has been recently ploughed. Water must be available at all times, either from a running stream, a trough or a bucket. Some form of shelter is necessary too, even if it is only by way of a hedge or a few trees to protect the animal from the wind and rain or to give it shade to discourage flies. A horse or pony living outside all year round must have some form of permanent shelter.

HOW BIG A PADDOCK DOES MY HORSE NEED?

A rough guide to a suitable area is a minimum of half a hectare (just over one acre) per horse; two acres per horse would be generous. Although horses are herd animals and enjoy each other's company, it is best not to keep too many together in a small area because it encourages worm infestations. In the summer it is possible to keep many more on the same area of grass, but monthly worm treatment of all the horses remains essential. Grazing should preferably be divided over two halves, and each half grazed alternately or rested, rather than grazing the whole area continually.

An ideal paddock has strong, secure fencing that is suitable for horses (ie not barbed wire), and shelter provided from the prevailing wind by a thick hedge *(left)*. Good pasture management – cutting weeds, harrowing, daily removal of droppings, grazing with cattle or sheep and not over grazing with horses – has resulted in a good, level sward of grass.

An unsuitable paddock has weak, unsafe or unsuitable fencing *(inset)*. Prolonged over-stocking has resulted in all the grass being eaten, forcing horses to eat the 'rough' areas around droppings which will give them heavy worm infestations. The ground has also been heavily 'poached' by the horses' feet in wet weather. This field will need ploughing, re-seeding, and re-fencing before it is suitable for horses.

IS IT ALL RIGHT FOR HORSES TO GRAZE WITH SHEEP, CATTLE OR DONKEYS?

Horses do not suffer from any infestation of the worms that infect sheep or cattle, and vice versa. It is therefore a good idea to graze a paddock alternately with horses and with other farm stock (except donkeys). This helps to cut down on the number of worms in the pasture, because horses can eat cattle worms and cattle can eat horse worms without either suffering any ill effects. Horses do occasionally pick up warbles (larvae) when grazing near cattle, however.

Horses are in some ways very poor users of land. They do not graze where they have urinated or defecated, and they tend to graze exclusively on the best areas of a paddock. Cattle and sheep graze more evenly, and can improve a paddock which horses have used by levelling up the grazing. Donkeys, on the other hand, commonly suffer from lungworms — although they may not show any symptoms. But horses that share a field with donkeys frequently develop a cough from having picked up lungworms, and it is therefore not advisable for horses and donkeys to share grazing.

WHAT IS THE BEST TYPE OF FENCING FOR HORSES?

Post-and-rail fencing is the ideal fencing material for horses, although it is expensive and horses tend to chew it or rub against it and push it down. Three rails are sufficient. The top should be 1.5 metres (4 feet) high, which should discourage a horse from jumping it. Uprights should be sunk 0.6 metres (2 feet) into the ground, and the top rail bolted — not nailed — to them. The bottom rail must be at least 30 centimetres (1 foot) above the ground to prevent an animal from getting its foot caught underneath it, and slightly lower for ponies and foals. To save expense, the middle rail can be replaced by single-strand high-tensile wire, or electric fencing. The top rail can alternatively be replaced by stud rail — a white PVC strip held under tension by two wires running through it. In fact, all three rails may be replaced by three strands of high-tensile wire or electric fencing, but if they are allowed to sag at all, animals may get their legs caught in them.

Post and rails need regular treating with creosote to stop them rotting (and to discourage horses from eating them). A healthy hedge is also a good fence — but horses tend to find gaps. It is often better to put fencing inside an existing hedge and to regard the hedge as shelter. For new paddocks it is worth considering planting a shelter belt of trees or shrubs outside the post-and-rail fencing.

It is important to go round the paddock and fence off any objects that a galloping horse might run into — for example electric or telephone-pole support cables.

Barbed wire, wire netting of any kind, and paling fences are totally unsuitable for use with horses.

The fencing illustrated on the left provides an example of unsuitable, bad fencing, whilst that on the right is both safe and secure. If there are any weak spots in the fence, a horse will quickly find them and push its way to freedom. Barbed wire or spike wooden paling fences can result in appalling injuries, and wire netting is not sufficiently strong to contain horses.

Paddock gates must be strong and securely latched. Ideally, they should not be situated in a corner of the field, as this can lead to animals being cornered and bullied when being turned out.

 ### WHAT ELSE DOES MY HORSE NEED IN A PADDOCK?

A gate is required, and should be at least 3 metres (10 feet) wide and as high as the fencing. It should not be situated in a corner of the paddock because horses being turned out there may be cornered and kicked by other bullying individuals. Do not build the gate in hollows or wet ground, or horses will congregate around it in winter, waiting to be fed, and make an awful mess. A few stone chippings in a gateway may alleviate such problems. If mains water is available, an automatic galvanized water trough is useful; if not, water will have to be brought in. A horse drinks an average of about 36 litres (8 gallons) a day. Wooden feed bowls can be fixed to the fence, but movable bowls for feeding on the ground are better and lessen 'poaching' of the ground if moved each day.

SHOULD A PADDOCK HAVE A SHELTER?

Horses or ponies at grass all through the winter need some form of shelter to give them protection from the weather. It may also help to keep their fodder dry and stop them wasting it. A shelter should be at least as large as a loose-box for a horse, and preferably a little larger; it should be open on the side facing away from the prevailing wind. The doors should not be too small, or bullying animals may prevent others from entering or may corner them inside. For this reason two doors, situated one at each end of a side, are preferable. A hard floor is essential, for a horse cannot rest when standing in cold mud. A little bedding will encourage an animal to lie down.

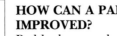 ### WHY ARE PASTURES SOMETIMES SAID TO BE 'HORSE SICK'?

When horses have grazed the same ground for many years, a very high level of contamination with worms can develop in the pasture, so that animals rapidly become infested and fail to thrive. For foals placed in such situations, this can be disastrous — fatal bowel damage can occur. This also explains why ponies kept on the small paddocks around large cities are often in such poor condition. The paddocks even look 'sick'. Much of the ground is eaten down to the bare earth; other areas have very long, lank grass that is full of weeds. The problem can be overcome with a change in management. If the situation is particularly bad it may be simpler to plough the ground, which should get rid of the worms, and reseed it with a grass mixture suitable for horses.

HOW CAN A PADDOCK BE IMPROVED?

Paddocks can be improved by resting them, by grazing with farm animals, by removing droppings, by cutting or spraying weeds, and by dressing with fertilizer. How much improvement is necessary depends on the stocking rate, and whether horses are permanently on the pasture or not. Harrowing with a chain harrow is very beneficial for paddocks: it not only breaks up and disperses piles of droppings, it also pulls out any dead grass. Specific agricultural sprays should be used for individual weeds, such as docks and thistles. Do not use general-purpose weed-killers. Sprays should be applied directly, using a knapsack spray, and the manufacturer's instructions in relation to keeping animals away from grazing following treatment should be strictly followed. Weeding, cutting in summer or topping long grass, produces better grazing in the autumn. Nitrogen fertilizers are unnecessary and produce too lush a growth of grass for horses. Lime, slag or potash may be needed to improve the soil. Many studs apply lime to their paddocks to increase the alkalinity of the soil and to stimulate the growth of plants that provide calcium for bone growth. Rolling and drainage also improve paddocks.

WHICH GRASSES ARE BEST — AND ARE THERE ANY PLANTS TO BE AVOIDED?

Unlike cattle, horses like the deep-rooted grasses that are found in permanent pasture. When reseeding a paddock, therefore, it is important to use a grass mixture recommended for horses. Consult your nearest agricultural advisory office for guidance. Horses like the most palatable grasses and perennial rye grass, timothy grass, meadow fescue and wild white clover all come into this category. Inevitably mixed with these grasses in a permanent pasture are less palatable grasses, such as cocksfoot, rough-stemmed meadow grass and tall fescue. Although horses enjoy clover, it can be very rich. Lime (calcium) encourages the growth of clover, whereas nitrogen fertilizers encourage grasses to grow. Horses eat many weeds, some of which, indeed, are beneficial. However, ragwort is very poisonous, causing chronic liver damage. Horses eat ragwort only if there is nothing else to eat, or if the ragwort plant has been cut and has wilted. The plants, with their yellow daisy-like flowers, are easily recognized in July and August and should be removed and burned whenever possible.

The grazing in this paddock is less than ideal for horses. The field is full of assorted weeds, so the amount of grass is very limited. The horses have eaten the grass right down, and there is virtually none left. Alternate grazing will have to be found or the horses will need supplementary feeding. Ideally, the weeds should be sprayed, the field cut and harrowed and then reseeded. It is when horses are kept in these kind of poor conditions that they look for weak spots in the fence to push through, in order to find more food.

 WHICH OTHER PLANTS ARE POISONOUS?

Fortunately, horses do not eat most poisonous plants — ragwort is the only common cause of trouble. Overhanging trees, such as yew — which is very poisonous — and laburnum, can be a problem; hedging shrubs such as privet are also poisonous. Some hedgerow and wood plants — for example, black and white bryony, or deadly or woody nightshade — and some plants that live in wet conditions near streams — such as hemlock and water dropwort — are other causes of poisoning, although cases are quite rare. Rather than eating fresh plants, horses more often eat hedge clippings thrown into the paddock. Acorns are poisonous if eaten in large numbers (this is a special problem in New Forest ponies), and bracken is poisonous if eaten continually over long periods. Rarely, a plant called St John's wort is eaten that contains a chemical which sensitizes areas of the skin to sunlight causing the skin to wither and slough.

ARE HORSES LIABLE TO SWALLOW POISONOUS CHEMICALS?

Horses have a very delicate sense of taste and smell, and seldom eat poisonous substances. All the same, they may very rarely suffer from lead poisoning through licking lead paint or old car batteries. Wood preservatives, such as creosote, may sometimes burn their lips or tongue when wood is chewed.

Organophosphoric poisoning, caused by a huge overdose of drugs used as wormers or insecticides, is rare. It can cause nervous signs such as twitching. Fungal toxins can also produce nervous symptoms; this type of poisoning occurs if food becomes contaminated with mould. Any mouldy concentrate or fodder, especially silage products, should be discarded and on no account be fed to horses.

STABLE ROUTINE

Horses are creatures of habit. A regular daily routine is important to ensure that they receive the correct food, exercise and attention, at the same time each day, to keep them well and in good condition. The daily care of a horse or pony is straightforward, although it may initially be difficult to decide between the wide variety of feeds, stable equipment, tack, and the other accessories that are available. The routine, once developed, will include a minimum of a daily inspection to check that the horse is healthy, and has adequate food, water and shelter when at grass. In the stable, keeping a horse healthy means regular mucking out and grooming, in addition to feeding and watering several times a day.

EQUIPMENT

Equipment for feeding and mucking out is essential for any stable, as well as a basic grooming kit.

WHAT STABLE EQUIPMENT DO I NEED?

The basic equipment for feeding, grooming and mucking out is essential for any stable. There must always be provision for fresh water, either in a bucket or from an automatic water drinker. Hay can be fed from either a haynet or a wall-mounted rack. Racks, however, are not recommended, particularly if they are mounted above head-level, for this makes the horse eat at an unnatural angle, and there is the danger that hay seeds may get in its eyes. Contrary to popular belief it is quite acceptable for hay to be fed on the ground, from where it is natural for a horse to eat, where the risk of dust and hay seeds is reduced, and the angle of which also assists the drainage of any nasal discharges. A permanent manger is

BASIC EQUIPMENT

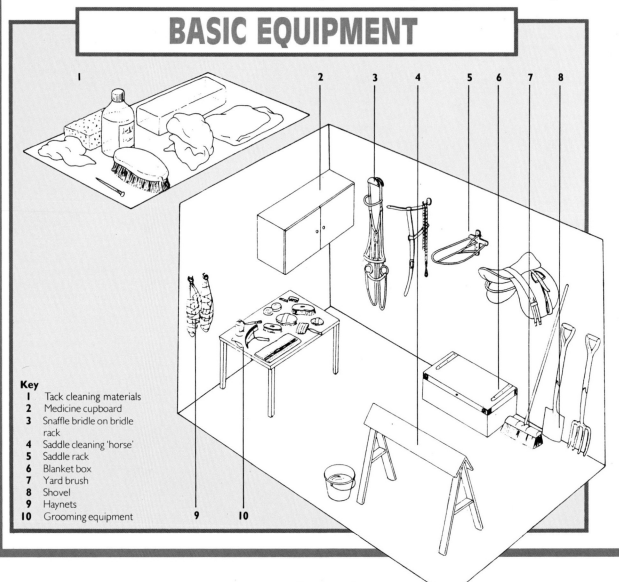

Key
1 Tack cleaning materials
2 Medicine cupboard
3 Snaffle bridle on bridle rack
4 Saddle cleaning 'horse'
5 Saddle rack
6 Blanket box
7 Yard brush
8 Shovel
9 Haynets
10 Grooming equipment

helpful for feeding hard feed and concentrates, but these can equally well be provided from a removable bowl on the ground, which can be made of plastic, metal or wood.

A basic grooming kit

This includes a hoof pick, a 'dandy' brush (with hard bristles), a body brush (with shorter, softer bristles), and a mane and tail comb. A curry comb is also helpful for removing excess hair and cleaning the body brush; and a plastic curry comb is particularly useful in dealing with muddy winter coats. Sponges are also essential: one for cleaning the nose and eyes, and another for cleaning the dock. Other grooming aids include a stable rubber (cloth), and a brush and pot for applying hoof oil. A 'wisp' of plaited straw or hay can be made on the spot for use in massaging the horse to promote circulation after grooming. Optional grooming equipment includes a sweat scraper and a water brush.

Equipment for mucking out

A broom, a shovel, and a four-pronged fork (which is preferable to a pitchfork) are essential. Droppings can be removed in a small container (called a skep): this may be a wooden basket, a metal or plastic bowl, or even a plastic laundry basket, which makes a very good substitute! A square of hessian sacking, or similar material, can be used to remove droppings or soiled shavings, sawdust or peat. A wheelbarrow is also needed to remove soiled bedding to the manure heap. A rake is a further extra required for peat or sawdust beddings. Finally, to clean and wash down the stable a pressure hose is preferable to buckets .

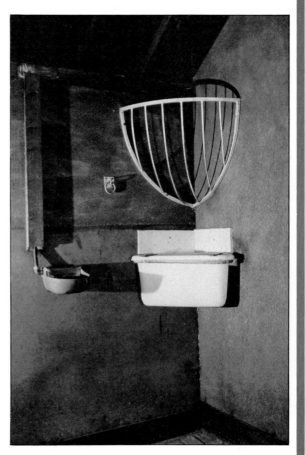

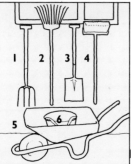

Equipment for mucking out

Keeping your horse's bed clean and fresh is a chore you must never neglect *(above)*. A major clean-out should be done regularly, as well as a daily removal of soiled bedding and droppings. The basic equipment you will need for mucking-out includes **1** a four-pronged fork, **2** a wooden or plastic rake, **3** a shovel **4** a broom, **5** a wheelbarrow or piece of sacking, **6** a plastic or wicker skep. **Stable fittings** include a hay rack, manger *(above right)*, ring for tying the horse , and an automatic water dispenser. The manger must be removed and cleaned after feeding, and the water dispenser should also be checked and cleaned regularly; bits of hay pulled from the net could soon block it.

THE STABLE ROUTINE

Horses need care and attention every day of their lives when kept in domestic situations. Time should be set apart for this, so that a routine can be established.

? WHAT IS THE NORMAL FEEDING PATTERN OF A HORSE AT GRASS?

Unlike cattle and sheep, both of which eat large amounts of food at a time and then ruminate afterwards, horses have relatively small stomachs in relation to their body size (in volume about a quarter of the capacity of a cow). They thus eat small amounts at a time. In the wild, of course, when grass is not plentiful this may mean that they nevertheless have to graze continually through the day. Under domestic conditions on better pasture, however, horses generally graze most in the mornings. The feeding pattern also tends to vary with the seasons. In summer, they normally graze throughout the day and rest during the night. In winter, it is too cold for them to rest at night; they keep moving and graze during this period in order to rest during the warmth of the day.

? HOW MANY TIMES A DAY SHOULD A HORSE BE FED?

Horses are unable to consume large amounts of food at any one time because of their relatively small stomach size. It is important, thus, that they have several small feeds throughout the day. At grass, this is no problem, and each horse can please itself. When stabled part or all of the time, hay should always be available, unless the animal is on a controlled diet. The amount of hard feed (corn or concentrates) required varies with the size of horse, the workload it has to manage, and the time of year. The amount given should be spread through three feeds a day — and possibly four feeds when a larger total quantity is required for a stabled horse in full work. Feeds should be given at the same time each day. For a horse in work, a small feed is usually given early in the morning (one and a half hours before exercise). The main meal is given after exercise, and a further meal in the late afternoon. A fourth meal can be given as the last thing at night, if this is appropriate.

Overfeeding of concentrates at any one time can cause colic. Likewise, bolting food can also result in choke or digestive upsets. Putting large stones or bricks in the manger may help to prevent this problem in particularly greedy individuals.

? HOW MUCH HAY DOES A HORSE OR PONY NEED?

The amount required depends on the size of the animal, the work it is doing, and the time of year; on whether the horse is stabled or grazing, and if the latter, on the quality of the grass. Hay is suitable only for providing a horse's maintenance rations, and hard feed is required to make up any extra necessary for work. In some cases — particularly for thin-skinned animals in winter — concentrates may be required also for ordinary maintenance. A rough guide for a 16-hand horse is 11.4 to 13.6 kg (25 to 30 lbs) of food a day. When the horse is in heavy work this can be made up of 4.5 to 6.4 kg (10 to 14 lbs) of concentrate and the balance in hay. For ponies, a rough guide for a 12.2-hand pony is 2.3 to 4.1 kg (5 to 9 lbs) of hay a day with some access to grass; for one of 14.2 hands, 3.6 to 5.4 kg (8 to 12 lbs) similarly. A stabled pony, or one on very poor pasture, should receive double that amount, plus concentrates.

The quality of the hay fed to a horse is important. How good it is depends on the type of grasses it contains, how mature the grasses were when the hay was made, and the conditions at haymaking. Good well-made meadow hay is the best for horses. Some idea of the quality of any sample can be gained from the smell, which should be sweet. Mouldy and dusty hay should be avoided: these may cause fungal infections and lung problems in animals that are allergic to them. New hay is not very digestible and may cause colic. It should not be given to horses for at least six months after haymaking, and preferably not for a year to 18 months. Feeding horses is an acquired skill: each animal should be fed according to its needs. In practice, horses can have more or less free access to hay, except before the morning

FEEDS AND FEEDING

Some of the many **feeds** available for horses and ponies are shown here:

1	Hay
2	Bran
3	Nuts
4	Flaked maize
5	Chaff with molasses meal
6	Barley
7	Oats
8	Flaked barley
9	Sugar beet

Although hay is the most widely used bulk food for general maintenance rations, other forms of preserved grass are widely available, equally good and often of less variable nutritional value. Sufficient fibre must be included in the diet as it is essential for efficient equine digestion.

The special needs imposed by growth, pregnancy, hard work and cold weather all put extra demands on a horse's system and must be met by adding both protein and energy rations (carbohydrate) to the maintenance ration. Although specially compounded 'complete' feeds, such as nuts and cubes, have made the task of feeding horses much simpler, it still remains an art, not a science.

exercise. The hay ration is usually divided, and the haynet filled three times a day — a smaller quantity provided early in the morning, and the largest in the evening. Wastage indicates that the horse is being given too much. It should be remembered that horses at grass in very dry summers need to be fed hay in addition if the grass becomes parched.

WHAT ARE CONCENTRATES (HARD FEED), AND WHY DO HORSES NEED THEM?

Although hay or grass is normally sufficient to supply the energy and protein needs for body maintenance, a horse may be unable to eat enough hay or grass to supply the extra needs of regular work. Energy and protein must be supplied in a more concentrated form of food — hence the term 'concentrates' or, because it is harder than hay or grass, 'hard feed'. Concentrates are also needed to supply extra requirements during pregnancy, lactation, or growth. For work, concentrates must be fed according to the amount of exercise the animal is doing. If too much is fed (particularly of

oats) and the animal is not getting enough exercise, it may become 'fresh' or 'nappy', and is likely to develop conditions such as lymphangitis or azoturia. All these diseases are of course covered in detail in Chapter 8.

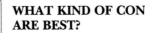

WHAT KIND OF CONCENTRATES ARE BEST?

Traditionally, oats have been the most frequently used hard feed for horses. When rolled or crushed, they are easily digested and have a high energy content (see table). Oats must not be stored for long after rolling — if not used within about three weeks, the nutritional value declines. Barley can also be fed, but is indigestible unless rolled or crushed. It is less likely than oats to cause horses and ponies to 'hot up'. Boiled barley (whole grains boiled for from four to six hours) is sometimes used for tired horses (after hunting), as a change from normal feed, or to tempt a 'shy' feeder. Wheat is unsuitable for horses (except as bran).

? HOW MUCH HARD FEED SHOULD I GIVE?

As a guide to quantity, a 14.2-hand pony, doing light hacking, needs between 0.9 and 1.8 kg (2 and 4 lbs) of oats a day, in divided feeds. With heavy work (such as hunting regularly, or competing in three-day events), this could be increased up to 5.4 kg (12 lbs) a day. For convenience, compound concentrates and coarse mixtures made up of various corn and protein ingredients, often with added minerals and vitamins, have been developed for different requirements. They are usually in the form of nuts or cubes, and have energy and protein contents specifically designed for the individual needs of different groups of horses (see table). For example, nuts for working horses and ponies have a much lower protein content than high-protein creep feed for foals (fed to them before weaning from a manger in so restricted an area that the dam is unable to reach it), which must contain the protein level needed for rapid body growth.

? WHAT ARE THE ADVANTAGES OF CUBES OR NUTS AND COURSE MIXTURES?

The chief advantage of these feeds for horses is that they contain a balanced ration, saving the owner the trouble of mixing a variety of food ingredients to create the balance. Equally advantageous is the fact that the nutritional value is constant in any specified compound. This makes feeding simpler and avoids problems with the considerable variation in nutritional value that may be found between different batches of hay, grain or other feeds.

Convenience is another great benefit for most horse-owners. Cubes, nuts or mixtures can be obtained in small amounts, can be stored easily, and there is little wastage. Products with protein and energy content designed for specific purposes — racehorse cubes, stud cubes, creep feed, horse and pony nuts, and many more — can be combined with hay or other fodder to provide a complete ration. Compound feeds are also available which contain dried fodder and provide a complete ration in themselves. With all cubes and nuts, the manufacturers' feeding recommendations must be followed closely.

? WHAT OTHER FEEDS ARE AVAILABLE FOR HORSES?

Keeping a horse in good condition, healthy and interested in its food, is an art, not a science.

Bran is commonly mixed with oats (1 part to 3 of oats). It has a high protein but poor energy content, is mildly laxative when wet, and is often given as a mash — boiling water is poured on to bran flakes and the mixture allowed to cool to blood heat before being fed — particularly to sick or tired horses, or as a means of administering medicine.

Small amounts of green food, root vegetables such as carrots, apples, peas or beans are often

Feeding hay in a haynet is the most least wasteful method of providing it *(left)*. If hay is placed on the floor in the stable, even in a corner, it can get mixed up with the bedding, so that either the horse does not get a full ration, or it will eat some of the bedding too. If hay is placed on the ground in a field, the horse will often spread it around with his feet, again wasting a large amount.

A secure and dry **feed room** to keep horses' rations is essential *(top right)*. Hard feed should be kept in bins with fitted lids that will keep out all vermin. If a feed room is attached to a field shelter, it is a good idea to keep it locked.

added to the rations of stabled horses in order to keep them interested in their feed. Fibre is essential for the normal functioning of horses' bowels. Chaff (chopped oat straw or hay) may be added in small amounts — up to 0.5 kg (1 lb) — to the diet of horses receiving large amounts of hard feed, to help digestion and to make them chew it properly.

Flaked maize and sugarbeet cubes are popular with owners of stabled horses who wish to keep them in good condition, yet want also to avoid the problem of their horses becoming too 'fresh' on oats. Flaked maize is high in energy, but must not be fed in large amounts; sugarbeet cubes must be soaked for 24 hours to avoid the danger of choke.

Molasses is sometimes added in small amounts to the diet of fussy eaters to keep them interested. Cod-liver oil helps to produce a healthy coat, but horses do not like its taste, and so it is usually incorporated with a sweetener in various commercial products. Linseed also promotes a glossy coat, and is given to show horses for this reason. It should *never* be fed raw, for it may contain a poison in this state; boiling and simmering overnight removes the danger, so that linseed can be fed as a mash mixed with other feeds.

CAN HORSES EAT SILAGE?

Because of the ever increasing cost, and the reduced availability of good-quality hay in small bales, there has recently been an increased demand for alternative types of fodder. Dried grass or lucerne is widely available; both have almost as good a nutritional value as the original material. Produced as a meal (which is then usually soaked in water) or cubes (which can be fed whole or soaked overnight), they must be introduced into a diet gradually, and are a much underestimated food for horses. Horses do eat silage very well, although it tends to make their droppings rather loose. Various vacuum-packed types of silage and haylage are commercially available and, although relatively expensive, are very good nutritionally. They have the advantage that they do not produce the fungal spores commonly found in hay — which means they are very useful as feed for horses suffering from mould allergy.

Horses also eat clamp silage well, but there have been some severe problems with 'big bale' silage used as feed, due to fatal infections caused by a soil contaminant, resulting in cases of botulism poisoning. For this reason, feeding 'big bale' silage to horses is not advisable.

? HOW MUCH FOOD SHOULD I GIVE MY HORSE?

Weighing out quantities of food is misleading because, unless a complete compound food is used, the weight required varies with the quality of the ingredients (particularly hay). In working out a ration two things must be borne in mind: the maintenance requirement, and any special requirements — that is, work, growth, pregnancy or lactation. The maintenance ration depends on body size and the breed and type. The other important factor is the time of year — how much of the fodder is to be supplied by grass. In summer, good grazing should supply the maintenance needs of all horses. In winter, however, both the availability and nutritional quality of grass decreases markedly, and additional fodder — by way of hay, dried grass or silage-type products — is required even for maintenance. Making up a maintenance ration for an animal that is always stabled is simpler, and does not require the seasonal balancing of feed for those stabled for part of the day only, or for those permanently at grass. As a guide, for an average pony of 14.2 hands, maintenance requirements when stabled and doing light hacking only would be 3.6 to 4.5 kg (8 to 10 lbs) of hay a day, plus 0.9 to 1.4 kg (2 to 3 lbs) of concentrate.

A bran mash is made by mixing bran with boiling water, stirring it well, then covering with a sack or cloth and leaving until it is cool enough to feed. Bran has virtually no nutritional value; fed dry (mixed with other grains) it can have a constipating effect, but when made into a mash it is useful as a mild laxative for horses that are somewhat off-colour or confined to their stable.

? WHAT EXTRA RATIONS ARE NEEDED?

Increased rations of fodder and, more particularly, concentrates are needed most commonly when horses are doing regular heavy exercise — the so-called 'working ration'. Growing horses also have greater feed requirements, and high-protein concentrates are given to foals and yearlings to assist body-building. Mares also require extra feeding during late pregnancy (the last three months) and during lactation (six months). Winter also puts an increased energy demand on animals, depending on where they are kept, for maintaining body warmth. Clipped and thin-skinned horses — especially Arabs and Thoroughbreds — require additional rations to cope with cold conditions. Native ponies and coarser animals do not normally need such supplements.

? ARE MINERAL AND VITAMIN SUPPLEMENTS NECESSARY?

A healthy body and a healthy coat both require adequate vitamins and minerals. There is no advantage in supplying more than these basic needs, however, and horses at grass do not require such supplements unless they are ill. When stabled, horses do not have access to vitamins and minerals from grass and the soil, and these must accordingly be provided in the diet. In a balanced compound concentrate ration (nuts or cubes), they are normally incorporated in the correct amounts — and this is perhaps the most convenient way of overcoming the problem. If an owner is mixing his or her own feed, a commercial mineral and vitamin supplement should be given. The most important needs in this respect are for salt and calcium. Small amounts — teaspoonsful — of table salt can be added to the ration, and horses also appreciate salt-licks on the wall, or rock salt placed in the manger.

High-level performance, especially in hot weather, puts extra demands on a horse's constitution, particularly one that sweats profusely. Such animals may require more salt in their diet than does a normal working horse.

Calcium is especially important to growing animals, and is most conveniently supplied in the form of limestone. Small amounts administered daily ensure good bone growth in young animals.

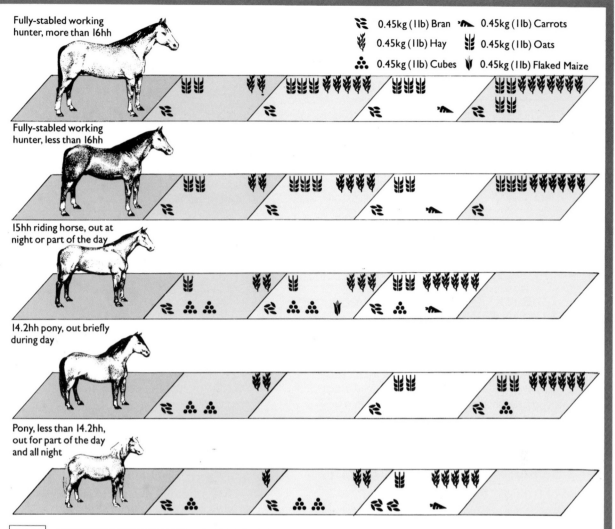

Fully-stabled working hunter, more than 16hh

Fully-stabled working hunter, less than 16hh

15hh riding horse, out at night or part of the day

14.2hh pony, out briefly during day

Pony, less than 14.2hh, out for part of the day and all night

0.45kg (1lb) Bran	0.45kg (1lb) Carrots
0.45kg (1lb) Hay	0.45kg (1lb) Oats
0.45kg (1lb) Cubes	0.45kg (1lb) Flaked Maize

HOW MUCH WATER DOES A HORSE NEED?

As a rule, horses should have unlimited access to clean, fresh water, either in the stable or out at grass. Horses obtain some of their fluid requirements from food, so the amount 'drunk' each day depends not only on the size of the horse, the air temperature and the amount of exercise, but also on the diet. Horses should not be given large amounts of water before or during exercise; it is better to water them before feeding and after exercise. After a very hard day's work, it is better to give smaller drinks (which have had the chill taken off them) from a bucket at 15-minute intervals, rather than allowing unrestricted amounts of cold water from an automatic feeder. A medium-sized horse drinks 36 to 45 litres (8 to 10 gallons) of water a day, and up to 68 litres (15 gallons) in hot weather.

Suggested **feeding guides** for horses and ponies kept in varying conditions *(above)*. Exact rations will still depend on the individual animal's make-up, and the work required.

If the stable has a **manger** as a permanent fixture *(below)*, feed can be given in this, but it should be cleaned out and scrubbed regularly.

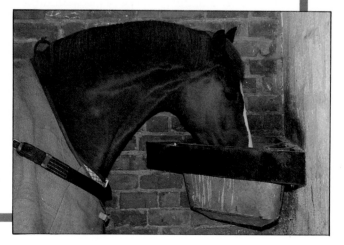

GROOMING

Grooming a horse has much more to it than merely improving the animal's appearance; it also helps to keep it healthy and feeling good. Stabled horses and those kept at grass have different grooming requirements.

 DOES DIRT OR MUD DAMAGE A HORSE'S SKIN?

Unaccountably, horses seem to like mud and take great pleasure in rolling in it! The reason they do so is unclear, but it presumably provides extra insulation and warmth on top of an already thick coat. Dry, caked mud seems not to do any harm on horses at grass. When mud once dried becomes wet, however, it irritates the skin, and chafing may cause dermatitis; it may in turn also allow bacteria (particularly *Dermatophilus*) to enter the skin and cause infection — this is the cause of 'rain scald' and 'mud fever'. Mud should be removed regularly from the quarters and legs of horses wintered at grass in wet conditions, to prevent this. As a general rule, it is better to allow mud and dirt to dry, and then to remove them by brushing, rather than to wash them off when still wet.

Continual washing of horses' legs removes the natural water-repelling oils from the coat; in cold, wet winter conditions, it may also cause irritation and encourage dermatitis, particularly at the heels and back of the pastern ('cracked heels'). This is a moist eczema that heals only with difficulty because the raw area is continually being re-opened each time the horse moves: prolonged use of ointments and bandaging may be necessary to resolve the condition.

Mud and dirt under tack also cause problems, and the minimum grooming should always include brushing the coat under the saddle, girth, and bridle before exercise. Again, sweat and dirt should be removed by brushing when dry — not immediately after exercise.

HOW OFTEN SHOULD HORSES BE GROOMED?

Regular grooming not only keeps a horse clean but massages its skin, improving the circulation and toning up the underlying muscles; it also improves the condition of the coat. Stabled horses, for their own wellbeing, should be groomed thoroughly at least once a day. Additional grooming ('quartering' or 'setting fair') may also be needed during the day, depending on the breed (to which the type of coat corresponds), and on the amount of exercise. Horses at grass normally do not need grooming, other than occasionally in winter to prevent their coats from becoming matted.

WHAT GROOMING IS NECESSARY?

A brief grooming before exercise ('quartering') is essential. This combines picking out the feet, cleaning the areas under tack, brushing the mane and tail straight (using body brush and water brush), and washing eyes, nose and dock with a moist sponge.

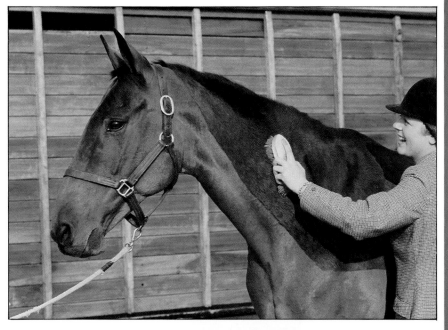

Regular grooming is essential for a horse's health and well-being. **The body brush,** used for brushing the coat of stabled horses, is pulled across the teeth of a metal curry comb to remove the dust and hairs *(left).*

Dandy brush, body brush and rubber curry comb *(top).*

The long coats of unclipped horses, or those out at grass, should be brushed with the dandy brush to remove any dried mud or sweat marks *(above).*

Brushing with the body brush helps to massage the skin and removes dust and grease from the coat *(right).* This brush should not be used on horses kept out at grass in the winter, as it will remove the grease they need to keep water out of the coat.

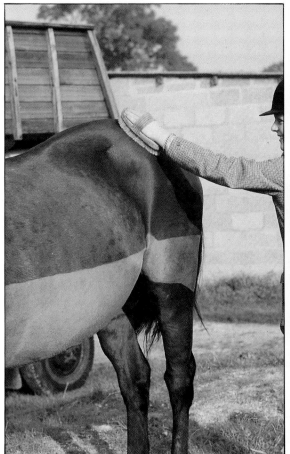

A thorough grooming ('strapping') can be given once the horse is dry after work. The feet are picked out first, beginning at the heels and working forwards to remove all mud from the sole and under the shoes, paying particular attention to cleaning out the clefts of the frog. Next, dried mud and sweat are removed, using the dandy brush or a plastic curry comb (these should not be used on the horse's head, which is sensitive and should be cleaned with a body brush). The hard work begins with the application of the body brush. This is used in a circular motion, following the direction of the hair, to brush the entire coat thoroughly. The brush is cleaned from time to time using a curry comb. Finally, a stable rubber is used all over the coat to remove the last traces of dust and to give it a shine. Good grooming may also include 'wisping' to massage the muscles. A hay or straw 'wisp' is used along the direction of the hair, with firm pressure over the muscled areas of the horse (that is, everything but the head, belly and lower limbs). A light grooming may be given later in the day when rugs are adjusted or put on — usually just a quick brush-over with a body brush.

? CAN I WASH OR SHAMPOO MY HORSE?

Washing the coat tends to remove many of the natural oils that help to keep the water out. Regular washing is not beneficial, although an occasional shampoo for a particular event or show does no harm. A specially medicated animal shampoo should always be used — not household detergent. Occasionally, specific dirty areas (like the flanks or the tail when soiled by droppings) can be washed using water and a cake of mild soap.

? HOW DO I KEEP MY HORSE'S MANE AND TAIL TIDY?

A mane comb should be sufficient to keep mane and tail unmatted, and should remove most of the loose hairs. If the horse has a very thick mane that will not lie flat, however, it may be necessary to thin it out: the longest hairs from underneath should be removed, one or two at a time, by plucking them between the fingers or wrapping them around the comb. Likewise, hairs may become too thick round the root of the tail, and become soiled by droppings: these too can be plucked — individually, beginning underneath and working sideways — to produce a tidy tail.

Tail bandages are sometimes worn to improve a tail's appearance; they are put on after wetting the tail with a water brush. They should *never* be left on overnight — if they are tight enough to stay on, they are too tight around the dock, and the 'tourniquet' effect can result in serious damage if the bandage is left on too long.

Thinning a pony's mane, by 'pulling' it *(above)*. Two or three hairs are removed at a time by wrapping them around a comb and giving a sharp tug.

? SHOULD THE TAIL BE CUT?

A horse's tail is useful for keeping the animal's flanks free from flies, and should be left long for this reason. To keep it tidy and yet effective, it should be cut off level with the point of the hock, when carried.

Some coarser individuals have large quantities of hair around the fetlocks. If these become particularly matted, removing the hairs altogether with scissors may help to keep the fetlocks clean. Special care should be taken while using clippers in this area not to cut the ergot (the small piece of horn at the back of the fetlock).

Washing a horse's tail helps to smarten the appearance for a special occasion, such as a show *(top)*. Care must be taken to rinse out all the soap.
The nostrils and area around the eyes should be carefully wiped with a clean, wrung-out sponge, at least once a day *(above)*. The sponge should be kept just for this purpose.

Hoof care. The correct way to hold a hind foot to pick out the underneath *(above)*. The leg is grasped around the pastern with the inner hand and held with the fetlock pressing on the person's thigh just above the knee. The outer hand is then used to clean the sole and frog using a hoof pick.

Applying **hoof oil** to the hoof wall *(below)*. This prevents it losing moisture and becoming dry and brittle, as well as smartening the appearance.

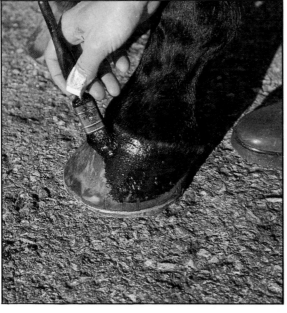

removed regularly, the moisture within it may penetrate the horn, making it soft and more liable to infection of the sole ('canker') or of the frog ('thrush'). Stabled animals should have all four feet picked out at least once a day. Feet should also be picked out before a ride, so that the animal has the maximum amount of grip, and afterwards to make sure it has not picked up any stones in its feet. Horses at grass must also occasionally have their feet picked out to remove clods of earth and stones, especially if they still have shoes on.

HOW CAN I KEEP MY HORSE'S HOOFS HEALTHY, AND PREVENT THEM FROM BECOMING BRITTLE OR FROM CRACKING?

Healthy hoofs must have a sufficient moisture content to keep them supple. If they become too dry they become brittle, flake, and may crack. This can lead to cracks in the hoof wall ('sandcracks' or 'quarter cracks') or to shoes being cast too easily. For horses in stables, regular dressing with hoof oil or grease prevents excess moisture loss and stops them becoming brittle. The oil or grease can help brittle feet to retain the moisture if applied after soaking them in water. Ponies' hoofs suffer from this trouble less often, and dressing may not be necessary. Individual animals may, on the other hand, have very brittle hoof horn, for which supplements of minerals (zinc) and amino-acids (biotin and methionine) may be of some help.

HOW OFTEN SHOULD FEET BE PICKED OUT?

It is helpful to accustom a horse (when young) to having its feet picked up. Approaching quietly, talking to the horse, and running a hand down the leg before picking it up, usually gets the animal to lift it up without being frightened. Matted earth at grass, or droppings, or bedding in the stable, can all collect under the hoof and shoe, especially around the frog. If such material is not

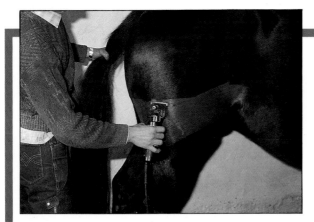

Clipping the horse's winter coat helps to prevent sweating and makes drying the coat easier. Trace clipping, shown here, removes a line of hair along the neck, under the belly and from the upper part of the legs. It is suitable for horses doing less strenuous exercise, or for those who are to be kept at grass. Clipping today is generally done with electric clippers.

Hunter clip. The legs are left unclipped to give protection *(above)*.

? DOES MY HORSE NEED TO BE CLIPPED?

A horse's coat serves the essential function of keeping the animal warm, and for this reason alters considerably between summer and winter. A major disadvantage of this is that a thick winter coat increases sweating during heavy exercise. Clipping is necessary only when horses undertake regular hard work, especially in winter. Usually, only part of the body is clipped. This reduces sweating and keeps the body clean.

? WHAT PATTERN OF CLIP IS BEST?

Various patterns of clip are shown in the illustrations. A full clip — in which all the hair is removed — is used on horses doing very heavy work (like racehorses). The saddle area and the legs can be left to provide protection from the saddle and brambles (the hunting clip). To keep a horse's back warm (and so avoid having to use several rugs) a blanket clip may be employed; this leaves the hair that would be covered by a blanket. The briefest clip is the trace clip (high or low, depending on whether the neck is included), which is useful in keeping the animal clean, since it removes hair from the areas where most sweat glands are situated.

? SHOULD MY HORSE BE RUGGED AFTER CLIPPING?

Skin thickness, which is related to the breed, makes a great deal of difference to how easy it is for a horse to keep warm. Native ponies and coarser types of horse can be partly clipped (low trace) without the need for rugs in the stable or, in mild weather, at grass. In severe conditions they obviously need the extra protection of a rug. Thin-skinned horses, when clipped, need a rug (and possibly an extra blanket) in the stable, and also require protection at grass — such as a New Zealand rug (a waterproof, canvas-covered rug). Horses fully clipped must be rugged all the time, except at exercise or in stables during the summer.

CLIPPING

Different types of **clips** *(below)*. The extent of any clipping should depend on the type of work the horse will do, and how much it sweats. The trace clip is the briefest and allows a horse to be kept outside, wearing a rug.

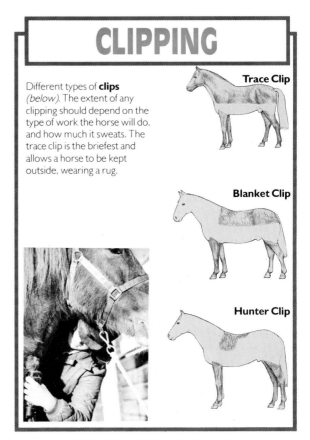

Trace Clip

Blanket Clip

Hunter Clip

TACK

There are many different types of saddle and bridle. For the novice, a general purpose saddle and a snaffle bridle are most suitable. They must be carefully fitted.

? WHAT TYPE OF SADDLE DO I NEED?
The type of saddle depends on the particular needs of the rider. Show jumping, eventing, dressage, showing, sidesaddle and racing — all have their own types of saddle, adapted to take into account the different features of each style of riding. For the novice a general-purpose saddle is best. Saddle size is measured from the pommel to the cantle, and standard sizes are 38 to 43 cm (15 to 17 inches).

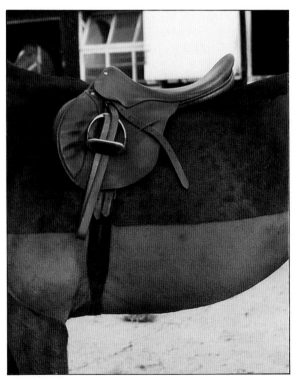

? HOW CAN I TELL IF MY SADDLE IS A GOOD FIT?
A good saddle should be well balanced: its weight should be evenly distributed, with no undue pressure on wither or back. The front arch must not be so narrow as to pinch the withers, or so wide as to allow it to press down on them; there must be no pressure on the horse's spine (to avoid pressure sores) and there should be a clear tunnel through the saddle when viewed from behind. The fit should be checked visually when someone is in the saddle. The advice of a reputable saddler is strongly advocated.

? WHAT KIND OF GIRTH IS BEST?
There are four main types of girth: webbing, leather, string, and nylon. Within these types there are other features to be considered — safety, ease of cleaning, and how easy it is to ensure that chafing does not occur. In the past, two webbing girths were commonly used, the second as a safety measure in case the first one broke. But webbing girths tend to rot, are hard to keep clean, and are seldom used today. Leather girths are probably the best, being strong, easy to clean with saddle soap, but need oiling from time to time in order to keep them supple. Nylon makes a good material for a general-purpose girth and is easy to clean. String girths are also satisfactory and less likely to cause galls than other forms (they are particularly useful on unclipped animals). A sheepskin or similarly padded sleeve is sometimes put around a girth as an extra protection against chafing or girth galls (which can be a problem in fat, unfit horses newly in work).

Many different types of saddle are available. A **jumping saddle**, illustrated *(left)*, has padded flaps, cut further forward than in most saddles. This allows for the knee to fit snugly against the saddle when the stirrup leathers are shortened. The deep seat helps keep the rider's seat and weight situated in the deepest part of the saddle.

A three-fold leather girth is very strong, but must be kept clean, otherwise sweat will dry and harden to rub against the horse's belly, causing girth galls *(right)*.

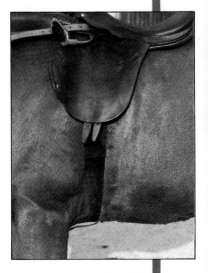

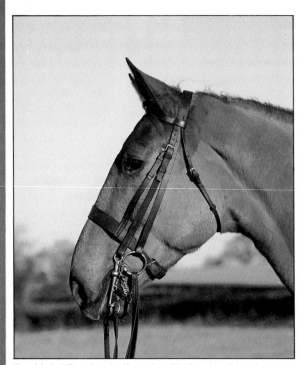

Double bridle – showing the curb and bridoon bits *(above)*.

Snaffle bridle with cavesson noseband *(above)*.

? WHAT KIND OF BRIDLE AND BIT DOES MY HORSE NEED?

There are three types of bridle: single, double and hackamore. The single bridle is used most often; the other types are used only by very experienced riders. The bridle enables the rider to control the horse by applying pressure and leverage variously to one, other, or a combination of the corners of the mouth, the tongue, the bars of the mouth (areas of gum in front of the cheek teeth), the front of the nose and the poll — depending on the type of bridle and bit (see diagram).

Bits can be made of any of a number of materials. Forged steel is best — and most expensive. Plated steel may chip; nickel and other softer alloys may wear and cut the mouth, and are unsuitable. Rubber and vulcanite bits are also used, and are softer than steel. The most common bit is the snaffle. This is used for the early education of young horses, and unless very fine control is required, or firmer control, most horses are ridden with one form or another of this bit.

In fact, there is a bewildering variety of snaffle bits. The simplest is a jointed snaffle with rings, which exerts its effect by leverage on the sides of the mouth. On some horses the rings tend to pinch the corners of the mouth, in which case a hinged ring is fixed to the jointed mouthpiece (to create an egg-butt snaffle), a commonly-used device to overcome this problem. Snaffles can have straight bars or be jointed; the severity of the bit depends on the material from which it is made, the thickness, the shape, and the means of jointing the mouthpiece. A German snaffle has a broad mouthpiece and is mild, compared with a racing snaffle which is narrow; a twisted snaffle, in which the steel looks as if it has been wound up from one end, is much more severe than vulcanite.

? WHAT OTHER BRIDLES ARE USED?

Double bridles give more precise control and are used in advanced equitation; they also allow greater control of head carriage and are used in showing and dressage. Double bridles should not be used on a horse or pony until the animals are fully accustomed to responding to a bit — that is, they have been well schooled in the use of a snaffle. The upper bit (the bridoon) is a thin snaffle; the lower (the curb) is H-shaped; and the combination makes a variety of instructions possible — by exerting pressure on the tongue, the bars of the mouth, the jaw (via the curb chain), and the poll (by means of leverage).

A 'Hartwell' pelham bit. This bit *(above)* can be used with two reins or a single rein attached to a leather strap that links the two rings on each side of the bit.

WHAT OTHER BITS ARE USED?

A Pelham bit is a variation of the curb bit that combines something of the effects of a curb and a snaffle in one bit. There are different types, which exert varying pressures on the bars of the mouth and, via the curb chain, on the chin groove. This bit can be used with two reins, but a leather loop between the bit rings (a Pelham converter) is sometimes employed to allow the use of a single rein — this lessens the effect of the bit in that it cannot act on both the corners of the mouth and the chin groove at the same time. For this reason, the bit is sometimes used on ponies that are too strong for their riders. A Kimblewick bit works on the same principle, and is used for the same purpose . . . although it is quite severe and must be used with caution. A hackamore bridle has no bit in the mouth; it exerts pressure on the nose, chin groove and poll. It is severe and is used only by expert riders on problem horses, and is not recommended for normal riding.

DOES MY HORSE NEED A NOSEBAND?

A cavesson is an almost universal addition to a bridle. It serves no useful purpose and is generally added purely to improve appearance — although it may be used to attach a standing martingale. Sometimes sheepskin is added to the front of the noseband to reduce a horse's vision and thus to keep its head down. A dropped noseband is used to stop a horse opening its mouth or crossing its jaw; it is narrower than a cavesson, and is fitted below a snaffle bit. It is sometimes used together with a cavesson noseband (in which case the combination is called a flash noseband), especially if a standing martingale is also needed. A crossed noseband (a grackle) has an additional strap running above the bit, and is used for the same purpose as a dropped noseband but is more effective.

DOES MY HORSE NEED A MARTINGALE?

Martingales are used to increase control of the head or to alter the pull of the reins. A running martingale has very little effect, but ensures that the pull of the reins comes from the correct direction, irrespective of what else the rider does with his or her hands. In addition, the neck strap is very useful for a novice rider to hang on to! A standing martingale runs from the girth to a cavesson noseband. It is helpful on a horse that throws its head about in the air, and is frequently supplied for novice riders of ponies in riding schools. Irish martingales are used to prevent the reins from going over the horse's head, and are not often necessary.

A standing martingale. This is an adjustable leather strap running up from the girth, between a horse's forelegs and attaching to a cavesson noseband *(left)*. It is held in position by a neck strap and is used to stop the horse throwing its head in the air, making it easier to control and less likely to hit the rider in the face with its head.

A crupper. A leather strap running around the horse's dock is attached to the back of the saddle *(above)*. This is especially useful for small ponies, to prevent the saddle slipping forwards or to one side.

MY SADDLE SLIPS: IS THERE ANYTHING THAT COULD STOP IT FROM DOING SO?

First check that the saddle is a good fit and is sufficiently padded. This is especially important on horses that have a poor conformation (low withers) or on those that have lost a lot of body condition. In either case, a numnah (a saddle-shaped pad of sheepskin, rubber, felt or padded material) or foam pad may help prevent the saddle from slipping sideways. On a very thin or narrow-chested horse there is often the problem that the saddle tends to slip backwards along the loins. A breastplate, consisting of a neck-strap passing to the girth (between the forelegs) and anchored to the D-rings on the front of the saddle, may help to hold it in place. Likewise, a breast girth — a webbing strap that runs around the front of the chest and is attached to both sides of the girth — may have the same effect. On small ponies, the saddle tends to slip forwards from the withers up the neck, particularly when the animal puts its head down to eat. An adjustable leather strap leading from the back of the saddle around the dock — a crupper — should overcome this problem.

WHAT ARE SADDLE GALLS?

A gall is a thickening of the skin resulting from pressure and friction from ill-fitting tack. Saddle galls are found on the withers and the middle of the back, usually on top of the raised vertebral processes of the backbone. Sometimes galls may occur on the side of the backbone because the gullet of the saddle is not wide enough and the saddle is pinching. In every case, removing the source of the pressure is essential. A numnah or a foam pad that has had the relevant area above the pressure point cut out should ease the pressure and enable the skin to heal.

WHAT CAN I DO TO HELP A SORE MOUTH?

First check that the bit is of the correct width — a jointed snaffle (when pulled straight in the mouth) should protrude 0.5 cm ($\frac{1}{4}$ inch) on either side. The cheek straps should also be checked to make sure that these are adjusted properly to keep the bit in the correct position — just touching the corners of the mouth. Some individuals have very soft skin at the corners of the mouth: for them the mildest possible snaffle bit should be used and, if the lips are cut, an ointment should at once be employed to aid healing. Cracked lips do not heal easily, and tend to crack repeatedly unless treated. Pressure from the rings or the sidebars of a bit can be prevented by using rubber bit-stops.

A common cause of horses' 'hanging' or failing to take the bit is dental problems. Sore cheeks result if the bit is pulled on to sharp cheek teeth. And of course horses develop a sore mouth if too severe a bit is used — it should be stressed that horses often pull if a bit is too harsh, and a horse that is difficult to control because it pulls should be tried with a milder bit rather than with a more severe one.

WHAT OTHER CLOTHING, OR ACCESSORIES, MIGHT MY HORSE NEED?

Rugs are used to keep horses warm in the stable, especially when clipped. A horse blanket fitted underneath the rug provides additional warmth when required. Such blankets may be kept in place by a roller of leather or webbing, which protects the spine from pressure, and is preferable to a surcingle (a plain strap), which may cause rubbing. Waterproof, canvas-covered New Zealand rugs protect a horse at grass from the weather, and any of a variety of anti-sweat rugs and sheets may be used after exercise.

Protective clothing is important for travelling. Many different types of boots are also available to protect horses from injury in accidents or from faults in their own action. Boots like this include brushing, speedicut, coronet, knee, hock, overreach, tendon and travelling boots.

EXERCISING

Exercising is an important part of a horse's life and, if it is kept stabled, an essential part. The amount and the kind of exercise needed will depend on the breed of horse, the work required from it, and the conditions in which it is kept.

HOW MUCH EXERCISE DOES A HORSE NEED?

The mechanics of a horse's bloodflow, especially in the feet and lower limbs, require regular movement for efficient circulation. Horses confined to stables for long periods tend to suffer from swelling of the legs through filling with fluid (oedema) due to poor circulation. As a minimum form of exercise, some daily walking is required: 15 to 20 minutes is sufficient. When riding, the amount of exercise must be governed by how fit the horse is, and in order to attain fitness where possible a regular daily exercise routine should be established. Getting a horse fit (in hard condition) requires a gradually increasing exercise programme. Once the horse is fully fit, reassessment of the amount of exercise required is common sense. A horse that has worked hard one day does not need two hours' work the next: a 20-minute walk to take the stiffness out of the joints or any swelling out of the legs is all that is necessary. Keeping a horse in hard condition requires about two hours' exercise daily, either led or ridden. To keep the back and girth regions hard, the horse ought to be saddled and ridden regularly.

DO HORSES SUFFER IF THEY ARE NOT EXERCISED?

Both circulation and digestion are improved by exercise. Animals left in stables tend to get constipated (impaction) and mild circulation problems (fluid-filled legs). Boredom accompanies lack of exercise, and many stable vices and other psychological problems may arise — weaving, crib-biting, wind-sucking and box walking may all occur following long periods of confinement.

LUNGING

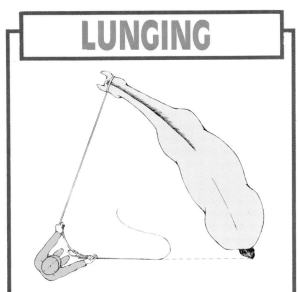

Training a horse to circle left on the lunge *(above)*. Positioned behind the point of the shoulder, the trainer gradually increases the size of the circle controlling the head with the left hand and the hindquarters with the whip in the right. The aim is to get the horse to walk, trot and halt correctly, responding to verbal commands.

The cavesson has three rings attached to the padded noseband *(above)*. The lunge rein attaches to the centre ring. After lunging on a circle, training can progress to long reining, in which the trainer drives the horse forward from the ground, working at different paces. The long reins attach to the side rings of the cavesson.

? I CAN EXERCISE MY HORSE ONLY AT IRREGULAR INTERVALS — IS THIS LIKELY TO CAUSE PROBLEMS?

Many horse-owners have difficulties in providing regular exercise, and find that their animals tend to be worked at weekends and not during the week. As long as they appreciate that the horse is unfit and do not overwork it, such a regime in itself is not harmful. However, owners should also be aware of the problems that can arise from irregular exercise. These are associated mainly with incorrect feeding for the amount of work that the animal is doing. The commonest problem is a muscular condition known as azoturia or tying-up (sometimes alternatively known as set-fast), which results when a horse that is fit and that has been receiving regular exercise is left in but given a full ration. The condition used to be common in working horses (carthorses and vanners) and was once also known as Monday-morning disease — because of the time it used to be discovered! Conversely, muscle damage can of course also follow overexercise of an unfit horse. Another problem that can arise from overfeeding and failure to exercise a horse is lymphangitis — inflammation of the lymph vessels — a circulatory disorder in which one or both hind legs may become greatly enlarged.

? HOW CAN I GET MY HORSE FIT?

A fitness programme requires regular work. When getting a horse up from grass, two weeks' very light exercise — walking — constitutes a good beginning. Continue with trotting and light cantering for the same sort of exercise periods during the next two or three weeks, increasing both the length and the amount of work regularly. It should take about six weeks to get the animal reasonably fit. For competitions in which peak fitness is needed, long exercise sessions are needed daily. Such sessions should include long hacking rides, road work and other forms of exercise to build up stamina, interspersed with periods of fast work (such as galloping or jumping).

CAN LUNGEING HARM HORSES?

Lungeing is a controlled way of exercising a horse without riding it, by getting it to run in a circle; a lungeing cavesson is used. The ring in which the exercise takes place should be as large as possible and have a good ground surface. Skill is required to lunge a horse properly; it can be quite severe exercise and so should be increased only gradually — but then consistently practised. Horses should not be lunged for long periods after days of comparative inaction. A tight circle also puts an additional strain on the limbs — animals with defective conformation (with legs that tend inwards or outwards) may suffer injury. In skilled hands, a carefully monitored lungeing programme can help build up a horse's strength and increase its suppleness. It is also useful when a horse is very 'fresh' and difficult to ride, or has a sore back.

Breaking roller. This is a wide leather strap that buckles round the horse's belly, and to which side reins are attached. A strap attached to the back of it runs along the horse's back and under its tail, to keep the roller in position. This is used when breaking-in a horse. The rings on the side of the roller are for long reins.

Catching a horse out at grass can be a job for two people if the horse is particularly fond of its freedom. It is best to take a bucket with a few nuts in it to act as a bribe. Horses and ponies that are notoriously hard to catch should be turned out wearing a headcollar.

WHY IS SWIMMING SOMETIMES USED FOR EXERCISING HORSES?

Horses swim quite well, and this form of exercise is occasionally used to give them a change when they have become sour or lost interest in work. Swimming involves great exertion for a horse's hindquarters, heart and lungs and is a useful way of exercising these organs and keeping a horse fit when it is lame or injured.

An equine **swimming pool** is beyond the means of most horse owners, but swimming *(left)* is a common feature of race horse training programmes today.

DO HORSES SUFFER FROM TRAVEL SICKNESS?

Horses are unable to vomit because of the anatomy of their digestive system, and in any case do not suffer from travel sickness. Some individuals are nervous travellers, however, and it may be necessary to sedate them to prevent injury. Rarely, horses on very long journeys (particularly on aircraft) may suffer transit tetany. This shows as involuntary muscle contractions and results from calcium deficiency. The exact biochemical mechanisms involved are unknown.

WHAT FOOD AND WATER ARE NECESSARY FOR A JOURNEY?

On short trips neither is necessary. However, a haynet distracts the animal and may make it travel more easily. On long journeys (lasting two hours or more) it is best to stop and offer the horse water before proceeding.

WHAT IS THE BEST WAY TO LOAD A HORSE INTO A HORSEBOX?

It is a great advantage if a horse is used to the box in which it is to travel. Time spent quietly accustoming a horse to enter a box, and feeding it there while the box is stationary, is time well spent. A horse that is accustomed to loading can usually be led up easily, using a head collar. If difficulty is anticipated, a bridle is essential for the greater control that it gives. It is best to allow plenty of time to coax the animal in, using titbits, encouragement and other inducements, before resorting to force. A lungeing rein can be an asset when loading without help — by attaching the rein to the offside of the bit or head collar, and around the hindquarters to encourage the horse forwards into the box. When help is available, two lungeing reins can be attached to the sides of the box and crossed behind the horse to encourage it forwards. Lifting each foreleg in turn and placing it higher up the ramp may persuade some reluctant

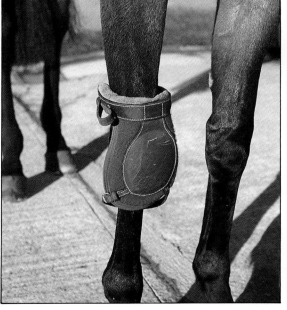

animals; someone encouraging from behind also helps. If this fails, the bristle end of a broom applied below the tail is sometimes enough surprise to achieve the required result! If this too fails, two people experienced with horses — not amateurs — linking hands round the hindquarters may be able to push the animal up the ramp. As a last resort, the animal can be blindfolded (with a cloth under the browband), turned, and led up. Here there is some danger that a horse may go off the side of the ramp and injure itself.

WHAT PROTECTIVE CLOTHING IS HELPFUL WHEN TRAVELLING?

In trailers, horses tend to lie back to gain support; this often causes damage to the root of the tail. A tail bandage is thus a routine precaution. Losing balance, horses may tread on themselves and damage their legs. These are normally protected, when travelling, by travelling boots (of foam) or travelling bandages over gamgee (cotton wool covered with gauze) which are applied from below the knee, or hock, to the coronet. For long journeys coronet boots can provide extra protection against treading. Padded hoods can be made to protect the heads of really fractious animals. To avoid the risk of injury, horses should *never* be tied up in a box until the bar, the breeching straps or the tail ramp has been put up.

Unloading at a show. Choose a quiet, preferably shady spot to park the horse box, and unload the horse in plenty of time *(left)*.

Headcollars are of many types and designs *(inset)*; they can be made of leather, hemp or nylon. Leather ones are probably the best, but, as with all leather tack, the leather must be treated regularly to keep it supple and prevent it from cracking.

A horse kitted out for **travelling** is seen here *(top)*. He is wearing a woollen travelling rug, held in place by a roller with breast plate and crupper attached. His forelegs are protected by thick stable bandages and knee-caps, and his hind legs by stable bandages and hock-boots .

Close-up of a **knee-cap**. This type is made of thick felt, reinforced with block leather directly over the knee and held in place by a strap fastening *(below)*.

Yorkshire boot: a thick cloth protective 'boot' for the fetlock, to prevent injury from brushing *(below right)* .

KEEPING YOUR HORSE HEALTHY

In addition to routine daily care, keeping a horse healthy requires other measures that must be carried out at regular intervals — like foot trimming, shoeing, and attention to teeth. Perhaps the most important contribution to be made towards maintaining health is proper worm treatment. Disease prevention is an essential part of an owner's responsibility, and tetanus and equine flu immunization should be routine precautions. It is helpful to compile a disease-control programme for your horse, and to enter the dates when the various procedures are due on a calendar or year-planner.

THE SIGNS OF GOOD HEALTH

Getting to know your horse and its usual behaviour will help you to judge when it is feeling off-colour.

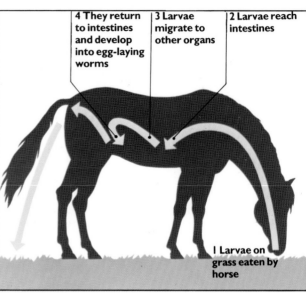

| 4 They return to intestines and develop into egg-laying worms | 3 Larvae migrate to other organs | 2 Larvae reach intestines |

1 Larvae on grass eaten by horse

The life cycle of the most damaging equine parasite – the large **redworm** Strongylus vulgaris *(above)*.

Horses always play host to a certain number of worms. It is very important to keep these under control with regular **worm doses** *(right)*. These can be obtained from the vet or saddler.

? HOW CAN I TELL IF MY HORSE IS IN GOOD HEALTH?

A healthy horse is bright, alert and in good body condition. Its coat should be flat and have a sheen (unless at grass in winter), and the skin should be supple, easily moved on top of the underlying tissues, and free from excessive scurf. The eye should be bright, and its lining membrane — and the membranes of the gums and those lining the nostrils — should be a salmon-pink colour. Horses normally stand during the daytime; if lying down when approached, however, they should get up quickly; they should stand evenly on all four feet, but may rest a hind leg (not a foreleg). Limbs must be free from heat or swelling. In motion, a horse's strides should be of equal length, the weight evenly distributed on all four legs.

The horse should eat well and chew its food properly. Frequent passing of droppings is important (about eight times a day); the faeces should be neither hard nor 'cowpat'-like, but should be a damp ball that just breaks up on contact with the ground. The colour of the droppings depends on the diet — in consistency they are looser when the animal is at grass. Urine is either colourless or pale yellow, fairly thick, and should be passed several times a day. The breathing should be even and regular, eight to twelve times per minute at rest, and can best be checked by watching the animal's ribcage movements from behind. The horse should not sweat at rest (except in exceptionally hot weather). The normal temperature is 35° Celsius (101.5° Fahrenheit). A normal resting pulse is 36 to 42 a minute.

? IS IT NECESSARY TO WORM MY HORSE, AND IF SO, HOW OFTEN?

All horses suffer from worms. They become infected when grazing by eating worm larvae hatched from eggs which are deposited in grass. Worms can live for long periods inside horses and it is therefore essential to worm all horses regularly, even those that are stabled. Horses pick up more worms during the summer when they are grazing and it is necessary to worm them comparatively frequently during this period. In the winter, worming is not so necessary because they are less likely to pick up worms. In addition to killing the worm and preventing damage to the horse, regular worming also helps to cut down pasture contamination. For effective worm control all horses in the same paddock should be wormed simultaneously. Every horse, including animals stabled all year round, should be wormed a minimum of twice a year; horses which live out at grass should be wormed every six to eight weeks, depending on the type of worm treatment used. It is particularly important to worm foals regularly because they have no immunity to worms. Worm damage at this age may cause problems for the rest of a horse's life, and may also stunt its growth. Foals should be wormed every four to six weeks.

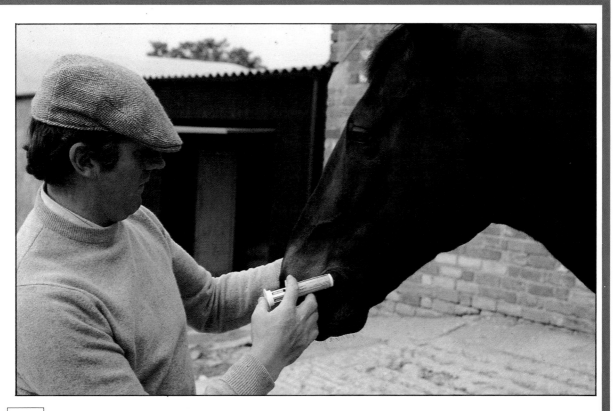

HOW DO WORMS DAMAGE HORSES?

Adult worms mostly live in a horse's intestines. There, they can live parasitically off the horse's food and digestive juices, and thus deprive the horse of nutrition. They also damage the lining of the intestines, causing them to become thickened, which makes it more difficult for the horse to absorb nutrients. When large numbers of worms are present, this alone can cause poor food conversion and poor condition.

The larval stages of development of many species of worm — and some adult worms too — invade a horse's body, where they damage its internal organs and cause a wide variety of problems. The most serious troubles, from a horse-owner's point of view, are caused by migrating larvae of the red worm, *Strongylus vulgaris*. These larvae leave the intestine and pass into the main blood vessels supplying the bowel, where they provoke a severe reaction and damage the walls of the blood vessels. This in turn may reduce the blood flow to the gut, or even block it off completely. Damage to the blood vessels from current or previous worm infestation causes pain in areas of bowel deprived of blood supply, and is the most common cause of colic in horses. Worm infestation in foals can have this effect years later.

Migrating worm larvae may also enter major blood vessels, causing blockages or dilatation (aneurysms). This happens most often in the anterior mesenteric artery (the main blood supply to the bowel), causing very severe colic. On rare occasions, larvae may invade other major blood vessels, causing fatal haemorrhage.

WHAT ARE THE SIGNS AND SYMPTOMS OF WORM INFESTATION?

Signs of worm infection are a poor condition, a dull coat, a pot belly, poor performance, anaemia and colic. Lungworm can also cause coughing. To assess the degree of infection and check a worm control programme, a worm egg count may be done. A sample of droppings is taken to a vet for analysis, and the number of worm eggs per gram of faeces is counted. This gives a good idea of the number of adult egg-producing worms in the horse. An acceptable level would be 200 eggs per gram or less. Recently, a blood test has been developed that can detect immature worms in horses; this can be used again to confirm or rule out worms as a cause of poor condition, and to check that worm treatment is being effective.

? WHAT WORMER SHOULD I USE?

A bewildering variety of worm treatments is available, and with every one it is important to follow the manufacturer's instructions carefully. Basically, wormers can be divided into the older type, which are cheaper, have to be given more often, and only kill adult worms, and newer wormers which are more expensive, can be used less often, and kill both immature and adult worms. Wormers are available as powders that can be mixed with food, as pastes for dosing by the owner, or in forms suitable only to be administered by a vet using a stomach tube. The choice between different kinds of products is often a matter of individual preference. There is evidence that worms can develop immunity to particular drugs. It is therefore important in any case to vary the type of drug used, and not to use one wormer continually.

? WHAT TYPES OF WORM AFFECT HORSES?

A wide variety of worms infect horses. Some are fairly harmless, and the animal can remain relatively well with large numbers of them in its body. Other species can cause serious damage. For instance, redworm infection can be fatal in young foals.

? ARE THERE ANY OTHER PARASITES AGAINST WHICH WORM TREATMENT MAY BE USEFUL — WHAT ABOUT BOTS, FOR EXAMPLE?

Adult botflies (*Gastrophilus intestinalis*) hatch in late summer but live only for a few days. During this period, the adult females hover about a horse and repeatedly dart at it to glue eggs to its hairs, causing the animal considerable irritation in the process. Large numbers of eggs are laid in succession, mainly around the fetlocks of the forelegs, but also higher up the legs and occasionally around the mouth. Those around the mouth hatch into larvae, which enter the mouth; simultaneously, eggs from the legs hatch within the mouth after they have been licked up by the horse. Within the mouth, the larvae penetrate its lining and migrate down to the stomach; there, they attach themselves to the stomach lining, and feed on the contents. They live as parasites in the stomach for 10-12 months before being passed in the droppings, pupating and hatching into adult flies. There is some argument about the amount of damage that bots do — some horses are found to have large numbers of bots in their stomachs, although showing no apparent ill effects. But bots do cause considerable ulceration of the stomach, and their presence in large numbers must affect the production of gastric juices and interfere with

Rubbing. Horses and ponies will often use fence or gate posts, or *(far left)* a handy telegraph pole, as a convenient place to alleviate an itch.

The result of such rubbing *(left)*. The cause of the irritation must be determined and treatment given accordingly.

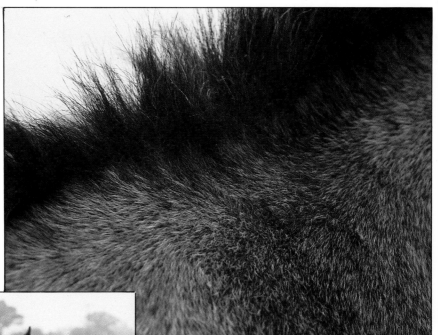

Sweet Itch. An allergic reaction to the bite of midges in the summer months causes intense irritation and self-inflicted injury to the mane *(above)* and tail *(right)*. The condition can be prevented by stabling sensitive animals during the evening when the midges are most active.

Lice are quite commonly found on horses and ponies in winter and spring *(left)*. They cause intense irritation and hair loss from rubbing around the head and neck, and, as in this case, on the hindquarters.

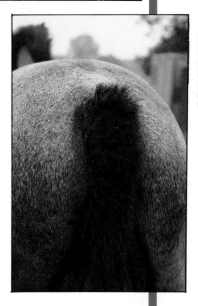

digestion. It is impossible to tell at sight whether or not a horse has bots in its stomach, and not all worm treatments are effective in removing them. It is a sensible precaution to use a wormer that is known to be effective against bots at least twice a year. Bot larvae, which are brown, pellet-like, and approximately 1 cm (just under half an inch) in length, can often be found in the droppings, showing that treatment has been effective.

Prevention is better achieved by removing the clumps of yellow eggs from the horse's legs in late summer through vigorous grooming with a hard dandy brush. If infestation is very heavy, it may be necessary to clip the legs to remove all the eggs. A large number of eggs that is repeatedly causing a problem may be destroyed by means of a weekly application of a 2 per cent carbolic dip to the parts of the animal on which the eggs are deposited.

PREVENTIVE MEASURES

Horses are very prone to catching infections from other horses, and many of these can be very debilitating. Protect your horse with vaccinations.

 IS FLU VACCINATION NECESSARY FOR MY HORSE? IF SO, WHEN SHOULD IT BE CARRIED OUT?

Equine influenza is a very infectious disease which can not only seriously affect the animal's health in the short term, but may cause permanent damage, resulting in lung, heart and nervous troubles. It occurs in epidemics and is spread from horse to horse by coughing. As with human flu, there are a variety of strains, and the immunity following vaccination or infection may not be 100% effective. However, very good protection can be achieved by vaccination, and in the few vaccinated animals that do contact the disease, the symptoms are much milder. Flu vaccination of foals can be carried out at three months of age. Vaccination at any age requires two initial doses, administered four to six weeks apart, followed by a third injection six months later. Thereafter, boosters must be given at intervals of a year or less. It may also be helpful to give an additional booster if a flu epidemic is present in the locality.

In an effort to reduce flu outbreaks, many organizers of competitions now require all entrants to be vaccinated against flu: Jockey Club, FEI (*Fédération Equestre Internationale*) and BHS (British Horse Society) events all require horses to be vaccinated. It is important to ensure that the vaccination certificate has been completed correctly, that injections have been given at the correct intervals, and that the horse is properly identified.

SHOULD MY HORSE BE VACCINATED AGAINST TETANUS?

Horses which for one reason or another have an open wound can be given temporary protection against tetanus by injecting them with tetanus antitoxin; such protection lasts for only three weeks, however. To provide full immunity, and to avoid the inconvenience and cost of repeated antitoxin injections every time the horse injures itself, it is strongly recommended that all horses and ponies be permanently vaccinated against the disease. Immunization cannot be given to foals under three months of age; at that age, though, two injections, given four to six weeks apart, are administered followed by a booster 12 months later. Thereafter, boosters every other year should be sufficient to maintain full protection.

WHAT INFECTIONS MIGHT MY HORSE CONTRACT FROM OTHER ANIMALS AWAY FROM HOME — AT A SHOW OR OUT HUNTING, FOR INSTANCE?

Respiratory diseases are the only conditions likely to be picked up in this way. There are several viruses that cause coughing and flu-like symptoms in horses. The most common is equine rhinopneumonitis (Equine Herpes Virus Type 1) which causes a watery nasal discharge, coughing, and high temperature. The horse is not usually very ill, and the infection commonly recurs because there is little immunity afterwards. No effective vaccination is available at present. Equine influenza is more serious, and horses are often ill for long periods afterwards. Other viruses (*Adenovirus* and *Reovirus*) and bacterial infections (strangles and other Streptococcal infections) can also be acquired from other horses and cause respiratory symptoms.

Preventing serious diseases such as tetanus and equine influenza by **vaccination** should be a routine precaution for all horses *(left)*. It is also important that booster injections are given at the correct intervals to maintain good immunity.

Girth galls are caused by rubbing from the girth, resulting in skin damage and infection *(right)*. They can be prevented by regular brushing of the coat to remove dried sweat and mud, and by keeping the girth clean and supple. If a horse's skin is very sensitive, a soft sleeve fitted over the girth will help to protect it.

ARE THERE ANY SKIN DISEASES WHICH CAN BE PREVENTED?

Horses suffer from many different types of skin disease, several of which can be prevented — especially those associated with ill-fitting tack (such as saddle and girth galls). Ringworm is quite common in horses, and is acquired from other horses via tack or grooming kit, or less often from cattle. It usually first appears on the parts of the skin in contact with tack, and is very infectious. Tack and grooming kit from an infected horse should be kept entirely separate and must be treated with a disinfectant that kills all fungal spores. (Ask a vet to recommend one because many common disinfectants are ineffective for this purpose.) Lice and mange can also be transmitted by grooming. As a rule it is not advisable to share grooming equipment between different horses. Grooming is important to remove excess mud from winter coats and thus prevent 'mud fever' and 'rain scald'. Cracked heels can be prevented by avoiding wetting (washing) legs in cold, wet weather.

WHAT SKIN PROBLEMS DO INSECTS CAUSE?

Insects are associated with many skin problems in horses. 'Sweet itch' is caused by an allergy to the bites of certain midges which attack horses around dusk in the summer months. This is a very tedious problem for owners, and one that can be prevented by taking stringent measures to stop the horse being bitten. These must include ensuring that the animal is in a stable between 4 p.m. and midnight, not permitting grazing near wet and marshy conditions, and using insect-repellents on the horse and fly-proof mesh and fly-killing devices in the stable (such as flypapers, Vapona and electrical devices). Nodular skin disease is also caused by an allergy to stable fly and horsefly bites, and fly-repellents and stabling can help reduce this problem. Horses grazing with cattle are sometimes attacked by warble flies, and migrating warble larvae cause problems when they emerge from the back the following year. Warble flies do not usually attack horses unaccompanied by other animals, so avoiding mixed grazing should prevent this problem. Warts are particularly common in horses under two years of age. They are caused by a virus, and normally clear up on their own without treatment. However, it is a sensible precaution to restrict contact with affected animals.

THE TEETH AND THEIR CARE

The horse's teeth are often neglected, but they can be the cause of digestive problems and give considerable pain.

? DO I NEED TO HAVE MY HORSE'S TEETH ATTENDED TO?

To digest food, a horse must be able to grind it properly, and needs level cheek teeth to do this. In addition, many equitation problems such as hanging to one side, failing to take the bit and, more rarely, head shaking, can result from teeth troubles. It is therefore advisable to keep a regular check on teeth. Unless a horse has specific dental problems, once a year should be sufficient, so it is often convenient to have the horse's teeth checked when your vet is giving it its annual vaccinations.

Dental problems are normally confined to the cheek teeth (premolars and molars). These should be worn level by the constant grinding action of eating. However, in practice this does not happen and sharp edges are formed, which cut the cheeks or the tongue. Sometimes upper cheek teeth are found closer to the front of the jaw than the lower ones; because there is no opposing wear, a 'hook' is then formed at the front of the first upper cheek tooth. This frequently causes a horse to hang to one side, or fail to take the bit properly, because the mouth is cut. Rasping is effective in removing these sharp edges, thereby preventing cuts; by levelling the teeth, the grinding of food is also made more efficient.

The first premolar tooth of domestic horses — the 'wolf' tooth — is often very small or absent altogether. Commonly they occur only in the upper jaw — but when present, they can be sharp and cut the cheeks, or may lie just below the gum surface, which becomes sore. Both these conditions cause problems with the bit. For this reason, wolf teeth are usually removed by a simple veterinary operation under local anaesthetic.

? WHY ARE TEETH PROBLEMS COMMON IN OLD AGE?

The teeth of very young and very old horses require more frequent attention. In young animals, emerging permanent teeth are very sharp when they first replace temporary (milk) teeth. As horses become older, the necessity for efficient grinding of fodder becomes more important to keep them in good condition. Uneven wear of the grinding surfaces is much more common in these animals. Teeth may have moved out of alignment, or may have been lost altogether, and the lack of wear from opposing teeth then gives rise to sharp prominences. Also, an irregular grinding surface allows food to accumulate in crevices, causing dental decay (caries). Older horses should have their teeth checked at least twice a year.

A yearling with soft, short, temporary (milk) incisor teeth *(far left)*.

Young horses will often experience teeth problems as their permanent cheek teeth replace the temporary ones. This takes place over the period from two to four years old.

Old horses have long, discoloured front teeth *(left)*. They are liable to suffer problems in their cheek teeth from irregular wear and dental decay.

TELLING THE AGE

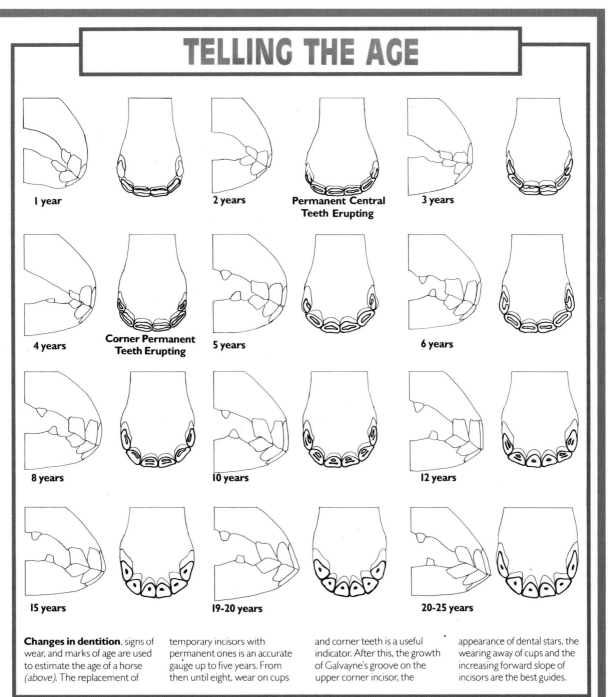

I year

2 years

Permanent Central Teeth Erupting

3 years

4 years

Corner Permanent Teeth Erupting

5 years

6 years

8 years

10 years

12 years

15 years

19-20 years

20-25 years

Changes in dentition, signs of wear, and marks of age are used to estimate the age of a horse *(above)*. The replacement of temporary incisors with permanent ones is an accurate gauge up to five years. From then until eight, wear on cups and corner teeth is a useful indicator. After this, the growth of Galvayne's groove on the upper corner incisor, the appearance of dental stars, the wearing away of cups and the increasing forward slope of incisors are the best guides.

IS IT POSSIBLE TO TELL HOW OLD A HORSE IS BY LOOKING AT ITS TEETH?

Horses can be aged with reasonable accuracy, up to the age of eight, by looking at their incisor teeth. Up to this age the teeth undergo clear changes each year. Beyond this, changes are less predictable. It is then only possible to give an approximate estimate of a horse's age, and horses are sometimes simply said to be 'aged'. The illustrations show the appearance of the 'table', or grinding surface, of the incisor teeth at various ages. When ageing a horse, it is important to distinguish temporary from permanent teeth.

FOOT CARE AND SHOEING

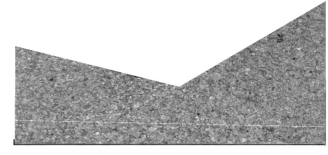

The feet are one of the horse's most sensitive areas and must be examined every day, whether the horse is stabled or at grass.

? IS IT NECESSARY FOR MY HORSE TO BE SHOD?

The horn of the hoof wall is formed at the top of the hoof from an area called the coronary band, or coronet. It grows at the rate of about 2.4 centimetres (one inch) a month, and it takes approximately six months for horn formed at the coronary band to reach the sole. In horses at grass, horn is worn away at the same rate at which it is formed, and shoeing is unnecessary. However, when a horse is exercised on hard surfaces (tarmac, concrete, and similar) horn is worn away faster than it is produced, and is often worn unevenly. Shoeing is therefore essential to protect the feet from excess wear and soreness, and even more vital for horses in regular work. Some horses at grass also need to be shod, especially in front, in order to prevent the hoof wall from splitting and breaking away. Where a sandcrack or quarter crack has begun, shoeing can prevent further cracking and allow the existing crack to grow out.

? HOW OFTEN SHOULD MY HORSE BE SHOD?

The rate of hoof growth varies slightly with individual horses, and with the time of year. On average, shoes should be removed every four to six weeks so that the extra hoof growth can be removed, the normal shape restored by paring, and the shoe renewed or refitted. Signs that reshoeing is necessary include cast or loose shoes, raised clinches and excessive wear, and misshapen feet. If shoes are left on far too long the outer hoof may grow around the shoe, particularly at the heels. This may push the heel of the shoe inwards, causing pressure and bruising of the sole, and is one cause of corns.

? WHAT IS A WELL-BALANCED FOOT?

The shape of a horse's feet can have a considerable effect on its action. Defects — especially long toes and low heels — put extra strain on the limbs and are a major contributory factor to the development of conditions like sprained flexor tendons and navicular disease. Good foot care involves keeping a horse's feet 'balanced' *whether they are shod or not*, this in turn involves regular trimming.

A well-balanced foot is one in which the weight is evenly distributed over the whole foot. The bearing (ground) surface should be level, and parallel with the coronet — that is, both sides of the wall should be the same height, and an imaginary line through the centre of the foot should pass through the centre of the pastern and cannon when viewed from the front. When viewed from behind, both heels should be the same height, the base of the frog just touching the ground. From a side view, the toe, quarters and heels should be proportioned. A balanced foot must have the front wall of the hoof parallel to the front of the pastern — a normal hoof-pastern axis. The hoof-pastern axis is very important; if it deviates from normal (is not parallel), alterations in the foot flight occur. An ideal front foot conformation is when the front wall of the hoof, the pastern and the slope of the heels are at an angle of 45 to 50 degrees to the horizontal. The hind feet are normally slightly steeper than the front, having a hoof-pastern axis between 50 and 55 degrees. If there is any doubt whether a foot is balanced, the length of the sides can be measured.

A very badly fitted **front shoe**, shown (*left*). The protruding edge is likely to clip the opposite foreleg, causing brushing injuries.

A case of **thrush**. The frog (*right*) has become soft and rotten by absorbing moisture from droppings and sodden bedding that has packed into the frog clefts. The commonest cause of thrush is the feet not being picked out regularly.

HOW SHOULD MY HORSE'S FEET BE TRIMMED?

Regular trimming of horse's feet is necessary, whether or not they are shod. At grass, trimming is needed to prevent the wall from becoming long and splitting. In addition, uneven wear — which results from defects in conformation or action — must be corrected. Horses with legs or toes that turn outwards thus wear the inner hoof wall more, and those that turn inwards wear the outside wall more. If the unworn wall is not reduced in order to level and balance the foot, the conformation defect may be compounded and made worse. Accordingly, horses that have legs or toes that turn outwards need the outside wall trimmed, whereas those with legs or toes that turn inwards need attention to the inner hoof wall.

Trimming is also required to maintain a correct hoof-pastern axis. This frequently involves trimming the heels, because the toes tend to be worn more at grass. When horses are fitted with shoes, the feet must be trimmed by the farrier every time the shoes are refitted (every four to six weeks). This process is known as 'dressing' the foot, and is a vital part of the farrier's craft. Horn tends to grow more quickly at the toe than at the heel. There is no wear at the toe when shoes are fitted. However, some wear occurs at the heels through friction resulting from heel expansion during movement — as much as 4.2 mm (one sixth of an inch) in a month. This is the reason shod feet become 'unbalanced', and extra strain is put on tendons as a result. The condition must be corrected by trimming to restore a correct hoof-pastern axis. Owners frequently do not appreciate how much can be achieved by a good farrier, by regular trimming. This is especially true in the case of foals and yearlings, in which major limb deformities can be cured by regular corrective trimming. In older horses response is slower, but repeated sessions of judicious corrective trimming can considerably improve very sloping or 'boxy' (upright) feet.

TREATING SANDCRACKS

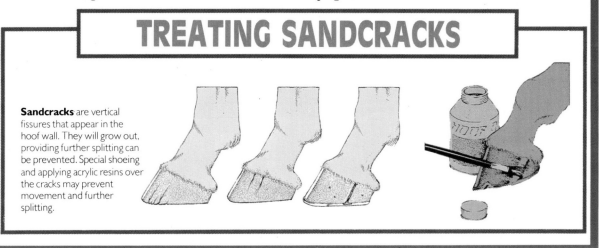

Sandcracks are vertical fissures that appear in the hoof wall. They will grow out, providing further splitting can be prevented. Special shoeing and applying acrylic resins over the cracks may prevent movement and further splitting.

BREEDING FOALING AND EARLY HANDLING

For horse-owners, the idea of breeding from their own mare has much appeal. The prospect of producing a foal with qualities similar to its mother, or even better, has many attractions. Before any decision to breed is taken, however, prior knowledge about normal breeding behaviour, what should happen at foaling, and how a newborn foal should behave and develop, is essential. For this reason, it is probably best for a novice to seek professional help with mating and foaling from a stud. Having once been responsible for looking after a pregnant mare and the care of her foal, it will then be easier in subsequent pregnancies to undertake more of the responsibility associated with this satisfying process.

THE DECISION TO BREED

Producing a foal from a much-loved mare is very exciting but there are many points to consider first.

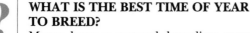

WHAT IS THE BEST TIME OF YEAR TO BREED?

Mares have a natural breeding season. Increasing daylight stimulates receptor centres in the brain, which in turn trigger the production of reproductive hormones. These hormones initiate the pattern of regular periods of 'heat', or oestrus, that characterize the breeding season each spring. These periods continue throughout the summer, and cease during the autumn. By artificially increasing the amount of light — for instance, by using electric lights in a stable — it is possible to begin the breeding season earlier. This practice is prevalent in Thoroughbred studs, which try to produce the foals as near as they can to January 1st — the official birthday of all Thoroughbred racehorses.

The ideal time for a foal to be born is between May and July, when most grass is available to help the mare's milk supply. Because pregnancy in horses lasts 11 months, the best time to have the mare covered is from June through to August — this is, in fact, the natural breeding time of Mountain and Moorland ponies in the wild.

SHOULD I BREED FROM MY MARE?

Although horses do not have many problems in breeding, it is advisable for novices to think twice before putting their mares in foal. Rearing a foal means extra work, and demands special facilities, including separate accommodation for the foal when it is weaned.

If the mare is pure-bred, there could be financial benefit from breeding. Breeding crossbred animals, however, is unlikely to be a financial success: the extra outlay required outweighs any potential profit.

Sometimes, mares that are retired because of injury are put in foal. Whereas this may help pay for the mare's keep, an objective view of her suitability for breeding must always be taken.

IS MY COLT GOOD ENOUGH TO USE AS A STALLION?

Very few home-bred colts are good enough to use as stallions. There are many first-rate stallions available commercially, and it is far better to use one of these. In any case, young colts are difficult to handle and are probably better gelded, unless there is a specific reason for not doing so. Both colts and stallions need expert handling, with the sort of experienced skill that is ordinarily available only on studs; it is difficult — if not actually dangerous — for amateurs, and is not to be recommended! Stallions that cover a mare only occasionally can also be very difficult to manage, and it is better to use one that is covering regularly.

COULD MY MARE BE TOO OLD OR TOO YOUNG TO BREED?

Mares often go on breeding until late in life, and suffer no ill effects from it. This is certainly true in general of animals that have bred regularly; it is more difficult to get an old mare in foal for the first time. Nevertheless, the only reason not to breed from an old mare would be that she had a medical condition which was likely to be made worse by pregnancy — severe lung problems, such as COPD (see Chapter 8), perhaps, or foot problems.

Fillies become sexually mature at around 18 months old, and can foal as two-year-olds. However, they are still growing at this age, and pregnancy may hinder their growth. Ideally, mares should not begin breeding until four years of age (to foal at five years), although some are put in foal when they are three.

HOW OFTEN DOES MY MARE COME IN SEASON?

During the breeding season from March to October mares show regular seasons lasting four to six days. These recur 14 to 16 days after the end of the previous season. Mares thus have a breeding cycle of around three weeks' duration. At the

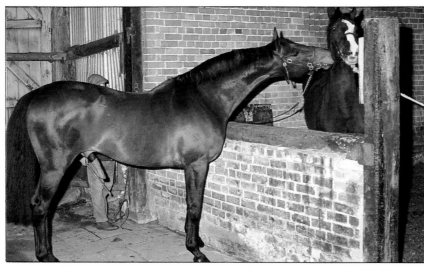

Teasing. When a mare has been sent to stud, she is frequently presented with a substitute stallion – usually of lesser value than the chosen sire – in order to test whether she is ready for mating. Should she not be ready, and kick or strike out as a result, it will matter less if she injures the substitute.

Covering. When the mare is ready for mating, she is presented to the chosen stallion. Careful thought and reseach is needed when choosing the right stallion for a mare, taking into consideration such things as temperament, conformation, action, constitution and soundness. It is also important to consider what is expected of the offspring – is it to be a competition horse of some kind, or merely a good, all-round family mount? Stud fees will vary enormously, running into huge sums of money for a famous and well known stallion.

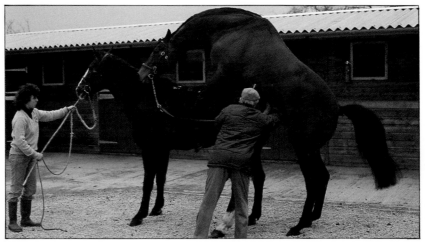

beginning and end of the stud season, mares may show irregular oestrus cycles. At any time, individual mares may return into season early, due to uterine infection. A foaling heat occurs four to eight days after birth. Mares can be covered then, if they are 'clean', or free from uterine infection.

WHEN WILL MY MARE ACCEPT THE STALLION?

Mares normally accept the stallion throughout the 'heat' period. It is, however, sometimes difficult to determine if a mare is truly in season. On studs, the willingness of a mare to stand for a stallion is often tested by using a stallion as a 'teaser'. When in season and confronted by a stallion, a mare normally stands, raises her tail,

urinates, and contracts her vulva. If she is not, she moves away and may kick out at the teaser. Mares normally ovulate 24 hours before the end of oestrus. For maximum fertility, they should be mated just before this time, not afterwards. On studs, veterinary examination of the ovaries, via the rectum, gives a good idea of when a mare is likely to ovulate. When such tests are not available, it is best to cover the mare on the second day of oestrus, and this should be repeated every second day until she goes out of season. At the start of the breeding season, mares may occasionally remain in heat for long periods without ovulating. This is particularly true of maidens, or mares which have not previously bred. There is little point in covering these animals, and veterinary advice should be sought.

SEXUAL PROBLEMS

If your mare shows signs of a sexual problem, advice should be sought from your vet — especially if you want to breed from her.

Inspecting the vulva of a pregnant mare can be helpful to see if foaling is imminent *(above)*. In some mares, it will slightly swell and lengthen in the final two or three days before foaling.

 MY MARE DOES NOT SEEM TO COME IN SEASON: WHAT CAN I DO ABOUT THIS?

During the breeding season, some mares may show no signs of oestrus, particularly if they are stabled. This is nothing to worry about, unless you intend to breed from the mare. Hormone treatments are now available that can either suppress or initiate oestrus in mares. Oestrus-suppressants are useful if a mare is very hard to manage when in season, but are not permitted for animals that are entered in competitions. It should also be remembered that pregnancy is of course a possible cause for mares not to come in season. Drugs used to induce oestrus cause abortion if a mare is already pregnant. Any possibility that the mare could be in foal should be thoroughly checked by a vet before such treatment is given.

MY MARE SEEMS ALWAYS TO BE IN SEASON OR TRIES TO MOUNT OTHER MARES: IS THIS ANYTHING TO WORRY ABOUT?

Although geldings may sometimes attempt to mount mares, other mares do not usually do so. Mares show prolonged oestrus early in the year, but if this occurs later in the season it may be a sign of a hormone problem in the ovaries. Some mares do not show oestrus behaviour and may also develop 'stallion-like' tendencies, attempting to mount other mares or exhibiting thickened neck or 'crest'. These symptoms may be caused by an ovarian tumour, and on their appearance veterinary advice should be sought.

MY MARE KEEPS COMING BACK INTO SEASON IN SPITE OF BEING COVERED BY A STALLION: WHAT IS WRONG?

This is usually a sign that the mare has a genital infection. Mares are normally 'swabbed' before they visit a stallion to make sure that no infection is present which may then infect the stallion and be transmitted to other mares. However, 'clean' mares may become infected at covering, and require treatment afterwards. Poor conformation of the vulva — if it is sloping, slack, or damaged at foaling — causes some mares to suck air into the vagina. This creates infection and prevents them from getting in foal. The vet can stitch the vulva to remedy this.

PREGNANCY PROBLEMS

You will probably need an expert opinion to determine whether a mare is pregnant. Feeding and daily care become particularly important when a mare is expecting a foal.

HOW CAN I TELL IF MY MARE IS IN FOAL?

Mares do not usually show much sign of abdominal enlargement until the last three months of pregnancy. The enlargement is obvious in narrow, light-framed animals, but in larger, broader-framed individuals it may be impossible to discern. Mammary development is usually obvious in maiden mares during the last month, but in mares that have previously foaled, 'bagging up' (mammary enlargement) may not be apparent until shortly before foaling.

HOW LONG DOES PREGNANCY LAST?

A normal pregnancy in horses lasts approximately 11 months — around 340 days. Colt foals tend to be carried longer than fillies. Premature foals may be born and survive, with intensive care, after 310 days' gestation. Foals may, on rare occasions, be carried three to four weeks over time.

HOW CAN MY MARE BE TESTED FOR PREGNANCY, AND WHAT IS THE BEST TIME TO DO THIS?

It is important to know whether a mare is in foal; ideally, she should not be ridden during the last five months of pregnancy. Pregnant mares also benefit from extra feeding, particularly in the last three months. There are many different ways of testing for pregnancy, some of which can be done only at certain stages (see table). Unlike females of other species, a pregnant mare produces a special

hormone in the uterus between day 45 and day 120 of pregnancy. Most owners like to have a blood test for the hormone carried out at this stage. Veterinary examination of the genital organs, via the rectum, can be used to check for pregnancy at any stage beyond day 40.

More recently, ultra-sound scanners are becoming increasingly popular for pregnancy diagnosis. By this means, pregnancy can be detected as early as 12 days after covering, but the scan is more often used from three weeks onwards.

A pregnant mare. Her greatly swollen sides are clearly visible from this head-on view *(below)*. This enlargement indicates the pregnancy is fairly far advanced; it is not usually visible until the last three months before the birth.

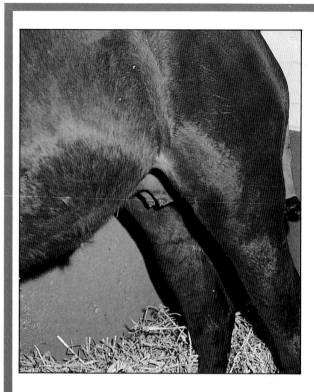

Signs that **foaling** is imminent *(left)*. The mare's udder is 'tight' and drips a thick, yellow wax-like secretion.

Early warning system for foaling *(above)*. The electrical alarm in the surcingle sounds in response to the mare's sweating, just before first-stage labour.

? WHAT ARE THE SIGNS THAT INDICATE MY MARE IS ABOUT TO FOAL?

Mammary enlargement is the most reliable sign of imminent foaling. A thick, waxy secretion may be seen dripping from the teats, and the mare is said to be 'waxed up'. Mares show few other signs that they are about to foal, and seem to be able to exercise some control over the process — often foaling at night and usually when there is no one watching! Slackening of the vulva and, less often, relaxation of the pelvic muscles and ligaments may be detectable, also indicating that birth is imminent.

? ARE THERE ANY PRECAUTIONS I SHOULD TAKE ON MY MARE'S BEHALF DURING HER PREGNANCY?

The largest part of a foal's development within its dam's uterus occurs within the last three months of pregnancy. Mares can safely be ridden for the first six months of the gestation period, but after this there may be a risk of losing the unborn foal. Extra food, particularly protein, is required during the last three months of pregnancy, to support foetal growth. Occasionally, mares may lose a pregnancy very early on, at about 20 to 30 days, if they are starved at this time. Mares put in foal very early in the spring may thus need supplementary feeding then. Nursing foals should be weaned from pregnant mares during the seventh or eighth month of the dam's next pregnancy in order to give the mare a chance to 'pick up' in condition. Regular exercise is important for keeping a mare in good condition; pregnancy sometimes interferes with circulation in the hind legs. Extreme cold seems to have no adverse affects on foetal well-being, and mares should be turned out for daily exercise, rugged up if necessary, rather than confined to a stable.

Owners should also be aware that mares sometimes show signs of being in season when they are pregnant, usually at around days 40 to 60 of pregnancy. This is not abnormal, and a veterinary examination or blood test can confirm that she is still pregnant. A horse's placenta — the organ that connects the mother's uterus to the growing offspring — differs markedly from that of humans and of other farm animals. The actual connection of the supply of blood and nutrients between the foetus and the lining of the dam's uterus is not particularly close; any reduction of the contact between foetus and the uterus greatly affects the amount of nourishment that passes between them. This is why twins, which naturally each have a separate placenta, are usually weak and poorly developed if born alive, but are more often aborted in late pregnancy. Infection of the placenta, known as placentitis, similarly produces an underdeveloped foal or causes an abortion. If vaginal discharge is noted during pregnancy, therefore, veterinary help should be sought immediately.

FOALING

Whilst most native pony and cross-breed mares generally experience little difficulty in foaling, some people like to send their mares away to foal, as complications can arise. In any event, make sure expert help is close at hand.

WHY IS IT UNCOMMON TO SEE A FOAL BEING BORN?

Horses rarely have any difficulty giving birth. The process is extremely rapid, usually takes place at night, and the foal should be on its feet very soon afterwards. Mares seem to try to avoid foaling when there are people about, and often manage to do so unobserved, unless a 24-hour watch is maintained.

SHOULD I TRY TO HELP A FOALING MARE?

It is normally unnecessary to interfere or to attempt to help the mare in any way at foaling. During the first stage of labour, the neck of the womb — the cervix — opens, as does the vagina and its entrance. The foetal membranes — the amniotic sac — which contain large amounts of fluid in which the foetus till now has been bathed (and which is sometimes called the 'water bladder'), may also be seen. The sac ruptures when powerful contractions begin during the second stage of labour. The contractions commonly cause the mare some degree of discomfort: at such a time mares are often restless, circle the box, paw the bedding, and may sweat up. They can foal standing up, but more often lie down to do so. The birth process is very quick. In most cases, the foal is presented and delivered forelegs first, closely followed by the head; the rest of the body appears soon afterwards, although mares may sometimes rest before expelling the hips with a final effort.

Second stage labour – the period of straining resulting in the birth of the foal *(below)*. It can last from 5-60 minutes, but averages around 20 minutes.

The foal pictured here is being born normally within its sac of foetal membranes, these will rupture during the final expulsive effort by the mare.

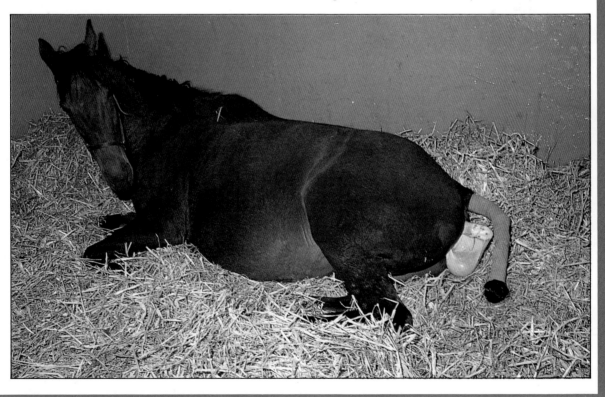

THE BIRTH

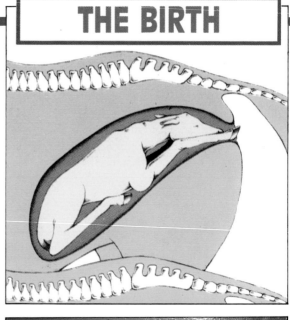

? HOW CAN I TELL IF THERE IS ANYTHING WRONG?

Knowing when something is going wrong is not easy, especially for an amateur. For this reason it may be better to send a mare to foal at a stud, where professional assistance is available. Nevertheless, even for an amateur it should be possible to tell from the way the feet are presented whether the foal is coming out forwards or backwards. If the mare continues to strain violently for a long time without making progress, a check can be made that all is well by inserting a hand into the vagina. Normally, the head can be felt directly behind the extended forelegs, lying on top of the knees. If all does not appear normal, expert help should be sought at once.

Sometimes the head may go back and halt the birth process. Foals are less often born backwards, but this usually presents no problems. In the rare case of a breech presentation, no legs appear, despite considerable staining, and the tail is felt when a hand is inserted. Veterinary help should be sought immediately, for it requires considerable expertise to realign the hind legs in a fashion ready for delivery. Deformities of foals, particularly contracted limbs, are one of the commoner causes of foaling problems. It is better to seek veterinary help soon in cases of trouble — delay is likely to result in the death of the foal. The person on the spot, however, should be aware that foals can suffocate if any part of the membranes remains round the nostrils after birth. In this case, the membrane must be cleared away by hand to allow the foal to breathe, and to prevent it choking.

Here the assistant gives the mare a helping hand, gently easing the foal's feet forward *(above)*.

? SHOULD I BREAK THE UMBILICAL CORD, OR TIE IT?

The umbilical cord contains the large blood vessels which have been supplying blood and nutrients to, and ferrying waste products from, the foal before birth. After foaling, uterine contractions compress the placenta, and blood passes from it for the last time down the cord into the foal. It is normally unnecessary to cut the cord; the foal can be left still attached to the cord until it is broken naturally, either by the mare, or by the foal getting to its feet. The placenta is shed in the third stage of labour. If at that time the cord is still attached to the foal, it may then be helpful to sever it, 2.5 centimetres (one inch) from the body; a ligature must be tied round the stump, which should be dressed with iodine or antibiotic powder. Keeping the navel clean prevents infection.

REARING A FOAL

Immediately after birth, all the care a healthy foal needs will be provided by the mare. However, it must soon get used to human care and attention as well.

HOW SOON DOES A FOAL STAND AND SUCK?

A foal's first movements after birth are usually to shake its head and to shiver, to help maintain body temperature. Very soon afterwards, it begins making attempts to get to its feet. A normal foal should be on its feet within two hours of birth — one that does not should be viewed with concern. Once standing, a foal soon finds its mother and, after a few unsuccessful attempts, locates her teat and starts sucking; this should happen within four hours of birth.

Unlike many mammals within the womb, a foal receives no antibodies — which are essential for protection from infection — from its mother before birth. Foals obtain their initial immune protection by taking in their dams' first milk, or colostrum, which is thick and yellowish. This is best absorbed when a foal is less than six hours old. After this, changes in its intestines mean that the foal is less able to absorb antibodies, and by 12 to 18 hours after birth it is unable to absorb them at all. It is therefore vital that foals receive the colostrum before they are six to eight hours old. If they have not done so by then, the mare must be milked and colostrum given by hand-held bottle, or by means of a stomach tube by a vet.

Second stage labour. The foetal sac with the foal's forelegs inside appears *(right)* An assistant is near at hand to give the mare reassurance and assistance, should she require it.

The foal is standing within half an hour of its birth *(right)*, and is being licked clean by the mare. Newborn foals should always be standing by the time they are two hours old, and have suckled within four to six hours.

HOW CAN I TELL IF A FOAL IS GETTING ENOUGH MILK?

Foals take small amounts of milk at frequent intervals, day and night, resting in between. It is hard to estimate the amount a foal is receiving, and in any case mares vary considerably in their milk output. Thoroughbreds produce from four to eight litres (one to two gallons) a day. Lack of milk is rarely a problem early in a foal's life, but it may stunt growth later, requiring early weaning. Occasionally, foals become gross and top-heavy, even suffering from limb deformities, because their dams' milk is too rich. This too can be cured by weaning. The composition of mare's milk alters when they are in oestrus, and foals may scour (suffer from diarrhoea) at this time. Orphan foals can be reared successfully using a proprietary horse-milk substitute; this has a more suitable protein content than cow's milk, which can be used, but which may cause scouring.

FOALING

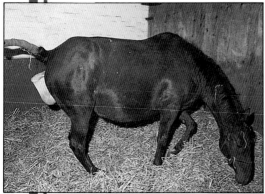

Mares frequently change position during second stage labour – that is, until the head is born, after which they usually remain in the same position. About five per cent of mares foal standing up; most lie stretched out on their sides to produce their offspring *(left)*.

This mare is experiencing problems *(right)*. To help her, the vet has **ruptured the foetal membranes** and is easing the feet forward.

The mare often ceases straining once the foal's hips are born, and she may lie for a long time with the foal's hind legs still in her vagina *(right)*. This period allows contraction of the uterus to squeeze blood from the foetal membranes down the umbilical cord to the foal. No attempt should be made to disturb mare or foal at this time, as this might sever the cord prematurely and deprive the foal of blood.

Dressing the navel *(left)*. Once the umbilical cord has ruptured it should be dressed with antibiotic powder to dry it and prevent any infection entering.
Within an hour or two of birth, the **new-born foal** should struggle to its feet. A gentle helping hand may be appreciated, whilst the dam nuzzles it encouragingly.

?

WHAT SIGNS OF ILL-HEALTH SHOULD I LOOK OUT FOR PARTICULARLY IN A YOUNG FOAL?

When a healthy, resting foal is approached, it should get up straight away and go to its dam and suck. Any foal that does not should be viewed with suspicion. 'Sleepiness' and refusal to suck — noticed because the dam has a tight udder — are signs that something is wrong; in such cases, the foal's temperature should be checked. The normal temperature of a foal is 38° Celsius (100.5 to 101° Fahrenheit); temperatures of 38.8 (102) degrees or more are significant, and if they are detected a vet should be called.

Bacterial infections are common in foals, especially if they have not received colostrum. Infection of joints causes lameness (joint-ill); infection of the lungs, pneumonia; and infection of the blood and thus the body in general, septicaemia. All of these infections cause symptoms of lethargy. Enteritis may also cause diarrhoea, and oral antibiotics may be needed.

Faeces are first formed in a foal's bowel while it is still in its dam. Called meconium, these faeces are hard and black, and usually passed in the first 12 hours after birth. Occasionally they remain in the foal, causing it great discomfort. An enema can prevent this, but if the foal is in pain, call a vet.

Tetanus can also occur in very young foals. On many studs they are given tetanus antitoxin for temporary protection until they can be vaccinated at three months old.

WHEN CAN A FOAL BE WEANED?

In the wild, foals are normally weaned at nine to 10 months if the mother has again become pregnant, but if not, they may go on sucking the teat for up to 18 months or two years. The quality and quantity of the mare's milk decreases after six months' lactation, and weaning is best done around six to eight months after birth, in the foal's first autumn.

Colts should definitely be weaned or separated from other females before they are one year old. Although they are not sexually mature until they are at least 18 months old, sexual behaviour begins a long time before this, and they may be kicked or injured by their dams.

Weaning involves considerable changes in a foal's digestion, and the more gradually the process can be carried out, the better. Creep feed, by which a foal is fed in an area so restricted that the mare cannot get to it, can be introduced when the foal is three to four months old and gradually increased until weaning is complete. Weaning is also a psychological upset for youngsters; some stable vices can be attributed to traumatic weaning. It is preferable to wean batches of mares and foals together, for this seems to cause less distress, and weaned foals benefit from companions.

HOW SOON CAN A HORSE BE HALTER-BROKEN AND LED?

Provided that a small enough head collar is available, foals can be led from the day they are born. The more foals are used to being handled the better: it can save a lot of trouble later on, and makes procedures such as worming, foot trimming and castration far simpler. Care should be taken in tying up young horses; tethering is not recommended for foals, who may panic. A quick-release safety knot should always be used.

WHEN IS THE BEST TIME TO BEGIN BREAKING-IN?

A horse's education can begin very early, and the sooner they are used to being groomed, having their feet picked out, and being handled generally, the better. If a young animal is used for showing, or regularly needs greater control than can be exercised with a head collar, it can be fitted with a bit as a yearling. However, it is not advisable to break horses for work until they are nearly mature. Thoroughbreds mature early, and are broken as yearlings, to race as two-year-olds. Eventers, hunters and hacks are not usually broken until at least three years of age, and more often as four-year-olds.

ARE THERE ANY PARTICULAR PROBLEMS THAT OCCUR DURING BREAKING-IN?

Although most horses are almost fully grown by four years old, they continue to develop and are not fully mature until they are six. Many of the problems encountered during breaking-in are caused by immaturity — extra stress on immature bone. In Thoroughbreds, this shows as sore shins or knee troubles, known as carpitis. In riding horses, splints on the foreleg cannon bones and, less commonly, curbs on the hocks are seen. Weakness in young horses may also lead to damage or injury, such as forging — the striking of the sole of the forefoot with the hindfoot — or brushing of the feet. Overwork can produce damage, particularly back injuries, which may not show up until later. Lungeing also puts extra strain on a young horse's body.

Sore and cut lips, cuts in the mouth, and saddle and girth sores are injuries directly related to breaking. The permanent cheek teeth are emerging at this stage of a horse's life (between two and four years) and these, together with growing wolf teeth, cause problems with the bit. Well-fitted rollers, saddles and girths are essential; the white hairs on many horses' withers are an indication that tack was not correctly fitted in the past.

BUYING A HORSE

Buying a horse or pony can be an absorbing challenge, but before setting out, the would-be owner must have a very clear idea of exactly what he or she requires — especially in terms of the approximate age and height, and the breed or type of animal. The major concern must always be whether the horse is right for the intended rider and the sort of riding he or she means to do. This will take into account his or her weight, strength and riding ability, and the funds, time, and facilities available for looking after the animal. There is no point in buying trouble, and it is a sensible precaution to have the horse 'vetted' before concluding any deal. It is worth taking the time and trouble to find a horse that is really suitable and liked by its new owner, because this is what the pleasure of owning a horse is all about.

THE CONSIDER–ATIONS

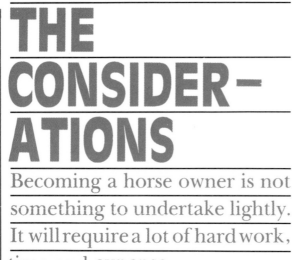

Becoming a horse owner is not something to undertake lightly. It will require a lot of hard work, time and expense.

long way away and the owner is working, particularly during winter. The minimum requirements are those of grazing and shelter, but if a horse is to be ridden frequently, some form of stable is essential. Keeping an animal stabled requires a minimum of one hour's work *every* day — for mucking out, grooming and feeding. This means that holidays or weekends away may have to be curtailed unless you can find someone else to do the work for you. If you do not have the facilities, a livery yard — offering either full livery or a do-it-yourself arrangement — can be considered. This saves the cost of buying stable equipment, and by sharing the workload with others, it may mean that horse ownership becomes feasible for someone who has other work to do.

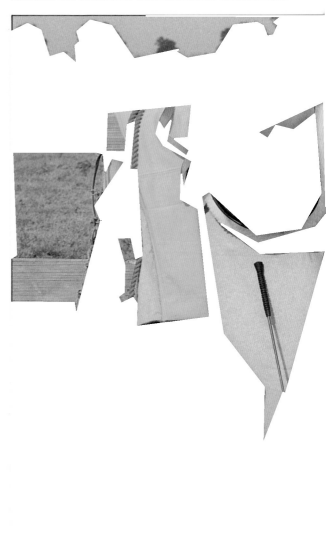

? CAN I AFFORD TO KEEP A HORSE?

Keeping a horse involves a considerable commitment, in terms both of time and of money. The initial expense of buying the animal is relatively small in proportion to the cost of keeping it (its feed, stabling, shoeing, medical care and so forth). Before buying, it is worthwhile working out approximately how much it all may cost, and how much time you have for riding. If funds and, more especially, time are limited, it may turn out to be preferable to hire a horse from a riding school when you wish to ride.

Having a horse on loan saves the initial expense but can also be fraught with problems, especially if the horse becomes injured or lame. Considering this, it is strongly advised that a written agreement be drawn up first, stating exactly who is to be responsible for specific costs (such as the veterinary fees in the case of accident or illness, or the expenses of vaccinations and insurance) and how, and under what circumstances, the animal is to be returned.

Tack is an extra expense. Although it can be bought second-hand, if it is in good condition it often proves to be nearly as expensive as new, and sometimes even more expensive! Well worn tack is a bad investment, and probably unsafe into the bargain.

? DO I HAVE THE TIME AND FACILITIES TO KEEP A HORSE?

Horses at grass must be checked over at least once a day; in stables at least twice a day. Such supervision may be difficult if the horse is kept a

❓ WHAT IS THE ABSOLUTE MINIMUM OF EQUIPMENT THAT MY HORSE WILL NEED ONCE I'VE BOUGHT IT?

For a young horse or a brood mare, the barest minimum must include a head collar, a leading rein, a basic grooming kit (hoof pick, dandy brush and body brush), and some feeding equipment (feed bowl, water bucket and haynet). For a riding horse, a snaffle bridle, saddle and girth are obviously necessary. A hard hat for the rider is a must, and the jockey-type skull cap (check that it conforms to current safety standards) is strongly recommended. Whether other stable equipment is necessary depends on whether the animal is to be kept at grass, in a stable, or at livery.

❓ WHAT OTHER EQUIPMENT IS USEFUL?

Additional equipment obviously depends on the type of horse, what it is used for, and where it is kept. If the animal is thin-skinned (an Arab, a Thoroughbred, or a cross from them) and to be kept at grass, a New Zealand rug is necessary. Leg bandages are useful for protection when travelling, to keep the animal warm, or for support after a hard day (in competition, hunting, or similar activity). Additional grooming kit may include a curry comb, tail comb, water brush and hoof-oil brush. Extra horse clothing — rugs, protective boots and so on — may also be needed. A buffer, hammer and pincers can be useful for removing loose shoes in an emergency, but are not essential. The extra length of a lungeing rein may occasionally be useful for leading a horse, and with a lungeing cavesson can of course be used for lungeing; it may also be a help when loading an awkward horse into a horse-box. A fire extinguisher is a necessary safety precaution for any stable.

❓ WHAT REASONS ARE THERE FOR CHOOSING BETWEEN A STALLION, A GELDING OR A MARE?

Whether a gelding is preferable to a mare is a moot point. Opinions differ, and it is really a matter of individual preference. If a gelding is allowed to grow before he is castrated (until two years old) it may be stronger, and have more muscular development, than a mare of similar parentage. Some people argue that mares are less reliable for riding because of their periodic 'seasons', and particularly for competitions. However, a mare's behaviour when in season varies greatly from one individual to another. Most mares cause few problems, and it is sometimes also argued that mares are more responsive. If a gelding becomes chronically lame, or permanently unfit for riding, its working life is over, whereas a mare could be used for breeding — provided that she is suitable, and that the time, skill and facilities are available. Stallions or colts require very firm handling by professional horsemen: they are not recommended for the inexperienced.

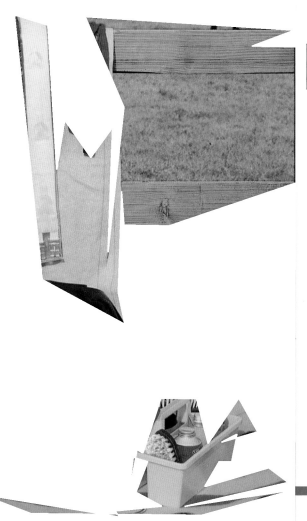

Some of the **basic equipment** needed for riding and looking after a horse *(left)*. Buying clothing, a protective hat and boots for the rider, and tack, protective clothing and rugs for the horse, as well as grooming, feeding, and watering equipment, entails a large capital outlay for the first time horse owner.

WHAT BREED OF HORSE SHOULD I LOOK FOR?

The majority of riding horses, eventers, show-jumpers, hunters, hacks, and so on are cross-bred animals. Often, they are the result of mating a cross-bred mare with a pure-bred stallion who, to some degree, stamps his offspring. Thoroughbreds and their crosses are bred for their speed and performance, and generally have light bones and frames; they are usually thin-skinned and require looking after during winter. Crosses with larger breeds, such as Hanoverians, produce the bigger-framed heavier-boned animals needed for carrying extra weight, or for competition. Native pony breeds are hardy, and suitable to live at grass. Ponies with Arab or Thoroughbred in their breeding may need cosseting during winter. There is little advantage in buying a pure-bred horse unless you intend to breed. However, if you wish to keep a pony out of doors all the year round, it is better to buy a pure-bred native pony (an Exmoor, Dartmoor, Welsh Mountain, Shetland or Fell), for the more pure these are, the hardier they are too.

WILL I OUTGROW MY HORSE?

'Child's pony sadly outgrown' is a common entry in the 'Horses For Sale' columns of newspapers. For a child it is preferable to buy an animal that is slightly larger than required; and for a child who is big for his or her age, a pony larger than that suggested by the rule of thumb quoted above will be needed. Although ponies can carry proportionately large weights, back and leg problems can result. It is better to part with an animal that is too small and buy one that is the right size, even after it has been a good and faithful servant.

WHAT SIZE OF HORSE DO I NEED TO CARRY MY WEIGHT?

The height of a horse is normally cited as an indication of its size, but the make-up of its bones and its frame must also be taken into account. A cob with a heavy frame and thick bones, for example, can carry more weight than a ladies' hunter of similar height. A rule of thumb to find a pony of suitable size for children is: a pony of 12.2 hands carries a child of 12 or less; of between 12.2 and 13.2 hands carries a child between 12 and 14 years; of 13.2 to 14.2 hands carries a rider up to 16 years. An animal of over 14.2 hands — technically

Height Measurement. The cross arm of the measuring stick is lowered onto the highest point of the withers to record the correct height of a horse *(above)*. The measurement is shown in 'hands' – each hand being 10cm (4 ins). A spirit level in the arm of the measuring stick ensures that it is level and the measurement accurate.

The kind of riding you will be doing, as well as the conditions in which you will be keeping the horse or pony, are important considerations when buying an animal *(right)*. These hardy little **Welsh ponies** can be kept out during even the harshest weather, provided they are kept well supplied with extra feeding rations.

this is a horse — is probably needed for anyone bigger. For adult riders, size depends on their weight; 15.2 hands is an average size for a ladies' hunter, and 16 hands and above for a man. Traditionally, horsemen have judged horses' bone by measuring the circumference of the foreleg cannon bone just below the knee. A hunter is said to have 'good bone' if the circumference at this point measures 21.6 cm ($8\frac{1}{2}$ inches) or more. This does not take into account the density of the bone, which is also important in weight-carrying. It may, however, be useful for comparing horses of similar height.

HOW DO YOU MEASURE A HORSE'S HEIGHT, AND IS IT IMPORTANT?

In spite of metrication, horses continue to be measured in hands — Shetland ponies, however, are often measured in inches. One hand is 10 cm (4 inches) — approximately the width of an adult human hand (including the thumb). The height is measured from the highest point of the withers to the ground. A measuring stick is used that has a sliding arm containing a spirit level; the

arm can be slid down the stick to the level of the withers to read off the height. For accuracy, certain conditions must be fulfilled: the horse must be standing on a smooth, hard, level surface; all four feet must be square; and the head should be lowered (so that the poll is level with the withers). The stick is placed behind the shoulder and forelegs, and the height at the withers is read.

Entry into various categories of competition — particularly showing and pony jumping events — is governed by a horse's height, and 12 millimetres (half an inch) can make the difference between eligibility and ineligibility for a class (such as the 14.2 hands and under).

To avoid having to measure every entrant for shows or competitions, a scheme for permanent certification of horses' heights has evolved. Under the Joint Measuring Scheme horses are measured, without shoes, by one of the officially appointed measurers. Certificates are issued for life, provided that the horse is more than six years old and has been measured by one of the officially appointed measurers. For a rough guide and for entry into Pony Club events, an allowance of 12 mm (half an inch) is usually made for shoes —

provided that the shoes worn are normal shoes, and not racing plates.

? WHEN BUYING A HORSE, WHAT IS THE BEST AGE FOR IT TO BE?

Generally, a horse is at its best between five and ten years of age, although its useful life may be anything between 20 and 30 years. Older animals are normally quieter, better schooled, and easier to manage. For a first pony, an old animal (over 15) is ideal: they are sometimes described as 'schoolmasters' because children can safely learn a lot from them. Young horses (under five years) may not be well schooled, may require experienced handling, and accordingly are not usually novice rides. A few wily horse dealers break in very young ponies (two to three years old) and try to pass them off as older animals. Being able to tell the age of a horse by looking at its teeth is a useful asset when going to look at horses for sale. However, it is not so easy for the novice, and it is better to get a vet or an experienced horseman to check for you if you are unsure.

WHAT ARE THE POINTS TO LOOK FOR IN AN ADVERTISEMENT?

The advertisement columns in local newspapers and horse magazines abound with horses for sale. Many of them have such marvellous attributes that you wonder how their owners can part with them! It is often what is left out of an advertisement, however — rather than what is said in one — that is important. Like the advertisements of estate agents, the equine equivalents have their own jargon. 'Genuine in every way', 'a generous horse', or a 'true christian' all mean that the animal has a good temperament and is co-operative. A 'patent-safety' or 'bomb-proof' pony is docile, and suitable for a nervous, or novice, rider. Such animals tend to be old and slow, and may be sluggish. Their virtue of safety may be due to the fact that they are too old, ill or lame to move other than slowly! It should be remembered that the quietest of horses or ponies can become naughty ('nappy') if consistently overfed, underworked and spoiled. 'Recently or lightly backed', 'broken this winter', 'ready for schooling', mean that the horse has done very little and is suitable only for an experienced rider. Likewise, 'well-broken' and 'green' mean that the animal has had little education. 'Not a novice ride' means precisely that, and may conceal a lunatic!

SHOULD I BUY FROM A DEALER?

Contrary to popular belief, buying through a horse dealer is often the best way of finding a suitable horse. A well-known local dealer usually has his reputation to think of, and is unlikely to try and pass you off with a dud. Having given him an idea of your requirements, and of what you are prepared to spend, he is usually able to come up with one or more suitable prospects. Wherever possible, it is advisable to have a horse or pony on a week's trial.

Buying at an auction is not to be recommended, particularly for the novice. There is neither the time nor the facilities to examine a horse thoroughly at an auction. There is also no chance to check up on the horse's behaviour (in traffic, in the stable) or any vices. An auction is a favourite place to dispose of an unmanageable horse, or one which has problems that can be temporarily disguised (such as sprained tendons). Warranties are sometimes given, verbally or in writing, testifying to an animal's age, soundness, freedom from vices, or usefulness for a purpose. These can be to the buyer's advantage if it should become

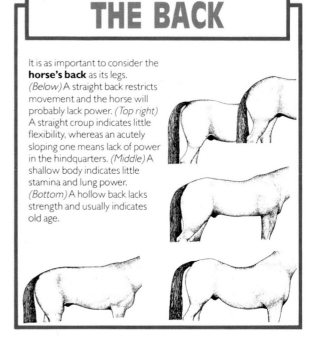

THE BACK

It is as important to consider the **horse's back** as its legs. *(Below)* A straight back restricts movement and the horse will probably lack power. *(Top right)* A straight croup indicates little flexibility, whereas an acutely sloping one means lack of power in the hindquarters. *(Middle)* A shallow body indicates little stamina and lung power. *(Bottom)* A hollow back lacks strength and usually indicates old age.

necessary to return the animal for failing to meet any of the warranted specifications. Statements such as 'believed sound' are merely an expression of an opinion, not a warranty, and are treated as such if it comes to legal action. Warranties concerning soundness are of less validity than those relating to age, freedom from vices or suitability. Particular attention should be paid to partial warranties of soundness — for example, 'warranted sound in heart, wind and action' could describe a blind horse! As a rule, leave auctions to the professionals. An apparent bargain could prove very costly, and there is often a good reason for such an animal to be sold at auction. There is usually no means of redress afterwards; *caveat emptor* — let the buyer beware — is not only sensible advice but the legal principle applied to public auctions.

IS ADVICE HELPFUL, WHEN BUYING A HORSE?

An 'eye for a horse' is the skill of looking at a horse's conformation and action, and being able to assess its potential and suitability for a particular purpose. This requires expertise which a novice cannot have. Another pair of eyes and a second opinion from an experienced horseman are always valuable when going to look at a horse for sale. In the end the buyer has to decide — but it is often a great help to be able to discuss the merits of the animal with someone else who has seen it, before making a decision.

WHAT SHOULD I LOOK FOR WHEN GOING TO SEE A HORSE THAT IS FOR SALE?

You will have to assess whether the animal is the right age, size, and type for your purpose; its conformation and action, and whether any faults are likely to predispose to problems. Signs of present or previous injury or illness will have to be taken into account, and it is essential to see the animal perform the sort of work for which you intend to use it. It is best to let its normal rider do this so that you can assess it from the ground first, and then try the animal for yourself. Finally, it is very important to assess the animal's temperament and how it behaves both in the stable and when ridden.·

HOW SHOULD I MAKE A GENERAL ASSESSMENT?

Conformation, or the 'make and shape' of a horse, as it is sometimes called, is an assessment of its physical characteristics and their relative proportions. Different breeds and types of animal have different conformations suited to their kind of work. In assessing conformation you must analyse the size and shape of the various parts of the body, their relative proportions, and how they are put together. You must then judge whether this is normal for the type of horse and, if not whether it is likely to put extra strain on the animal and so predispose it to illness or injury.

WHICH PHYSICAL CHARACTERISTICS SHOULD I CONSIDER?

The head is very important. Many people judge a horse's likely temperament and intelligence from its head. The eyes should be bright, large, and placed well to the side of the head. Small 'piggy' eyes placed narrowly together are often an indication of ill-temper. An elegant head is a sign of good breeding, and probably a hotter temperament. Coarser-looking animals are often more generous. The profile of the face should, be straight, and not convex (dished) or concave (Roman-nosed), although horses with the latter are often generous and genuine. Attention should be paid to the ears — whether they are pricked and alert, or are laid back when approached (a sign of ill-temper).

The neck must be in proportion to the rest of the horse. A short, thick neck can mean that the horse takes a strong hold, and may be awkward for the rider because there is very little in front of the saddle! Exceptionally long necks are weak, and

FRONT AND REAR VIEWS

Good conformation. Point of buttock is in line with hock and hoof.

A cow-hocked stance looks awkward, but is no problem if legs strong.

Good conformation. Point of shoulders is in line with knee and hoof.

Bow-legged conformation puts strain on hock bones and ligaments.

Pigeon-toed stance puts strain on the knees. The horse may tend to stumble.

Horse is **closed in front.** Has little heart room, may tend to brush.

make it difficult to achieve a good head carriage. Larynx problems (such as laryngeal hemiplegia) is more common in larger horses with long necks. Necks should be slightly arched (convex top line) and clearly demarcated from the shoulders. A straight neck can be improved by training, but a concave neck (ewe neck) is a permanent weakness. If the concavity is confined to the lower part of the neck only (swan or cock-throttled neck), it causes the horse to hold its head high, and, if excessive, makes for a very uncomfortable ride.

The spines of the backbone which form the withers should be large and prominent, so as to provide adequate anchorage for the neck ligament supporting the head. They should also be situated as far back along the horse's back as possible because, in this position, they are associated with long, sloping shoulders. Sloping shoulderblades are ideal, and give a horse large, sweeping strides. This makes the animal comfortable to ride, and this conformation is valued for both racing and jumping. Upright shoulders make for short strides, a jarring action, and an uncomfortable ride. However, they are associated with increased pulling power, and for this reason upright shoulders may be favoured in driving horses. The depth and width of the chest are important to provide plenty of room for heart and lungs.

WHAT IS THE IDEAL BACK CONFORMATION?

The back should be straight, neither arched upwards (roach back) nor downwards (dipped back). Short, muscular, broad, deep loins are preferable for both galloping and jumping. Long backs are weak, whereas short backs may make for an uncomfortable ride because the horse is prone to overreach. The angle of the hindquarters, or more particularly, the angle at which the pelvis joins the spine, affects the degree of thrust provided by the hind legs — in racehorses, for example, sprinters usually have more upright quarters than 'stayers'. Extremely upright quarters (goose-rumped) are often found in good jumpers; the top of the quarters is more prominent in these individuals which are said to have a 'jumper's bump.' The quarters should be rounded and well muscled, and wide, but excessive width may result in an uncomfortable 'rolling' action.

WHAT PROBLEMS ARISE THROUGH POOR LEG CONFORMATION?

Leg conformation is particularly important. It has a great bearing on the animal's action, and on any predisposition to lameness. In the foreleg, the shape of the knee is significant; it should be straight. If the knee is slightly forward ('over at the knee'), the effect is not very attractive but is unlikely to cause problems. Not so, however, if the knee is deviated backwards — 'back at the knee' or 'calf knees': this puts extra strain on tendons and ligaments, and is a serious fault.

The angle and length of the pasterns are important, and effect both the flexor tendons and the fetlock joints. The pasterns should be at an angle of 45 degrees when viewed from the side, and not too long. Excessively long and sloping pasterns put extra strain on the flexor tendons (and may cause strained tendons). Short, upright pasterns increase concussion and may contribute to arthritis of the fetlock joint.

When viewed from the front, a good foreleg conformation requires that an imaginary straight line from the point of the shoulder runs through the middle of the knee, fetlock and foot. Deviations from this put extra strain on different parts of the leg, and cause uneven weight distribution on the foot (with associated uneven wear of feet or shoes). The knees may be outside this line (bow legs) or inside ('knock knees'), and either defect puts extra strain on the relative parts of the knee joint. Horses' legs may turn inward or outwards below the fetlock, and this again puts extra strain on the fetlock joint; turning in ('pigeon toes') is worse than turning out ('splay-footed').

In narrow-chested horses, the feet are often wider apart than normal (outside the line from the point of the shoulder, or 'base wide'). This conformation results in wear of the inside of the foot or shoe, puts extra strain on the fetlock, and may cause arthritis with associated articular wind galls, although such animals are usually surefooted. Feet closer together than normal (inside the line from the point shoulder, or 'base narrow') are associated with broad muscular chests and result in excessive wear on the outside of the foot or shoe, and additional strain to the fetlock joint. This may also cause arthritis and articular wind galls.

WHAT IS THE IDEAL LIMB CONFORMATION?

In an ideal limb confrontation, a plumb line dropped from the point of the buttock (tuber

CHEST CONFORMATION

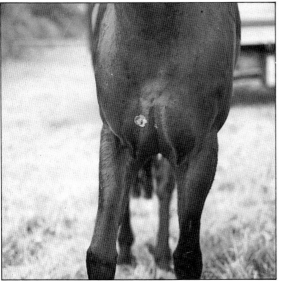

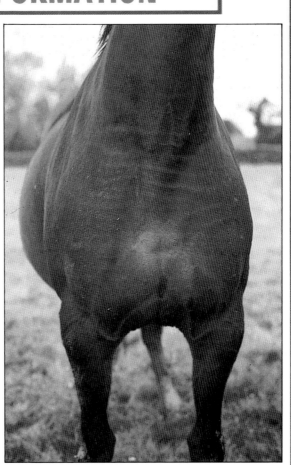

This horse *(above)* has a very **narrow chest** and will need to wear a breast plate or breast girth to prevent the saddle slipping back.

An example of a horse with a **good, broad and deep chest**, giving plenty of room for its heart and lungs *(right)*.

ischii) should pass through the middle of the hock, fetlock and foot when the horse is viewed from behind. When viewed from the side, a line from the same point (tuber ischii) should run to the point of the hock and straight down the back of the cannon bone to the ground. Animals with this ideal conformation are rare. Hind leg defects are more common and, fortunately, usually not as significant as those of the front leg. The hock, however, is a very important joint and does more work than any other in the body; it is responsible for absorbing most of the concussion in the hind limb. The hind leg should not be too straight or the horse is likely to prove a jarring, uncomfortable ride. Excessively curved or angular hocks are known as 'sickle' hocks: hocks like this are weak and put a heavy strain on the back of the hockjoint, especially on the plantar ligament which, when

sprained, causes a swelling known as a 'curb' to appear at the back of the hock. For this reason curved hocks are known also as 'curby hocks'. As with the forelimb, heavy muscling of the upper part of the body (in this case the quarters) is often associated with abnormal closeness of the feet ('base narrow'), often accompanied by hocks that are wider apart than normal (bow legs). This conformation causes strain on the outer aspects of the leg, which may show as such swellings around the hock as 'thorough-pins'. The opposite conformation, in which the feet are further apart than normal ('base wide') when viewed from behind, commonly results in the hocks being closer than normal and pointing towards one another ('cow hocks'). Such a disposition puts strain on the inside of the limb, particularly the hock joint, and is a cause of bone spavin disease.

FORELEGS

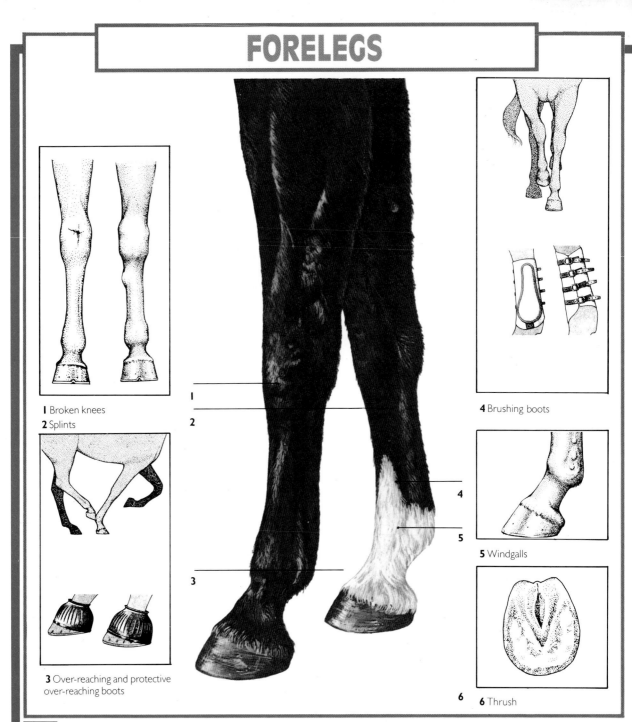

1 Broken knees
2 Splints

3 Over-reaching and protective over-reaching boots

4 Brushing boots

5 Windgalls

6 Thrush

HOW SHOULD I JUDGE FOOT CONFORMATION?

The conformation of a horse's feet is of supreme importance — the old horseman's adage 'no foot; no horse', says it all. The front and hind feet should each be a symmetrical pair: any differences between left and right should be viewed with suspicion. The feet should point straight forwards: deviations inwards or out affect the animal's action. Toes that turn out cause 'brushing', which may result in injury and necessitate protective boots, and are more significant than toes that turn in, which may cause 'dishing' and put extra strain on the fetlock joint. When viewed from the side, the front wall of the fore feet should be parallel to (and form a continuous line with) the front of the pastern — a normal hoof-pastern axis. Likewise, the heel

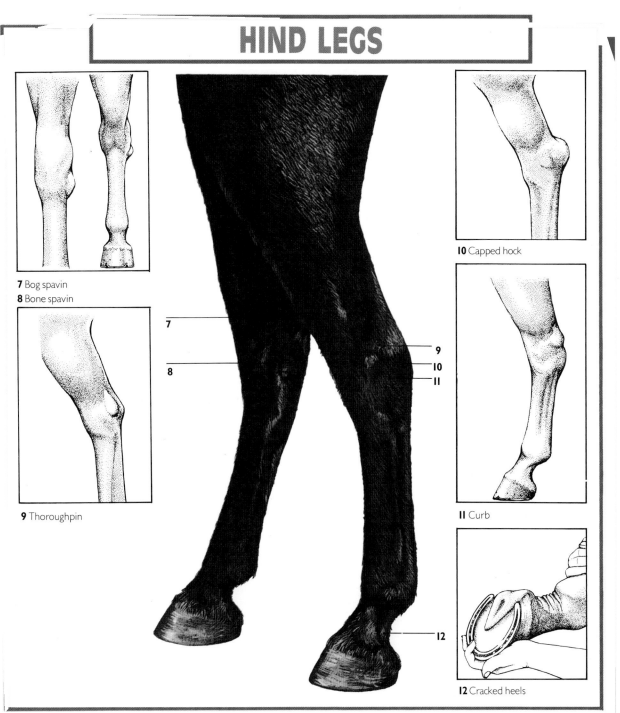

7 Bog spavin
8 Bone spavin

9 Thoroughpin

10 Capped hock

11 Curb

12 Cracked heels

should be parallel to the front wall of the hoof. The hoof-pastern axis is particularly important because it has a marked effect on the flight of the foot during motion — a sloping or upright hoof-pastern axis produces considerable alteration (although in many cases corrective foot trimming can remedy this). Excessively long toes and low heels are a major contributory factor in sprained tendons and navicular disease. Upright, 'boxy' (or 'donkey') feet have contracted heels and a smaller weight-bearing surface; this increases the effects of concussion, and should be avoided, for animals with this conformation are less likely to stand up to heavy work. The normal angle of the hind feet is slightly greater than that of the fore feet.

The wall of the hoof should be smooth, free from grooves, rings and cracks, and not dry or brittle.

Trotting a horse up – an essential part of pre-purchase examination by both prospective owners and their vet *(left)*. The trot is the best pace to detect lameness and faults in a horse's action. The horse should be trotted towards, past and away from the observer, on a loose rein, on a hard level surface.

? HOW IMPORTANT IS A HORSE'S ACTION?

A horse's action is important. It determines how well the animal is able to perform, and how comfortable the animal is to ride. If refined movement is demanded — in dressage for instance — a good action is a fundamental and vital prerequisite — whereas for an animal required only to do light hacking it is not so significant. Yet in all cases, a good, smoth, free, action makes for a comfortable ride. A short choppy action — one in which the animal does not flex its limbs properly — is uncomfortable for the rider and probably jars the horse as well. A horse's action depends on its conformation, and may be altered by lameness or injury.

To make an assessment of the way a horse moves, first watch it when it is walking: examine it as it walks away from you, directly towards you, and past you (that is, from one side).

? WHAT SHOULD I LOOK FOR IN THE PACES?

The trot is particularly important to assess action — most forms of lameness show up best at the trot. Although faults in conformation are generally noted when examining the horse at rest, some of them cause abnormalities in the animal's action, as can best be seen when the horse is trotted away or directly towards you. Thus, paddling — in which the foot moves forwards, inwards, and then outwards, in a circular motion — is associated with a 'toe-in' conformation , and wears the outer edge of the foot or shoe. Plaiting — in which the foot moves across and inwards to land more or less in front of the opposite foot — is seen in horses which a 'base narrow', 'toe-out' conformation, and is conducive to stumbling.

'Winging', or 'dishing' — in which the foot moves forwards, inwards, and then outwards in a circular motion — is common in horses with a 'toe-out' conformation, results in excessive wear of the inside of the foot or shoe, and is a cause of 'brushing'. A horse should move with relatively straight strides (when viewed from in front or from behind); swings and twists in action may cause extra strain on the joints, and a 'rolling' action makes an uncomfortable ride.

When a horse's action is viewed from the side, the feet should be seen to be picked up well and to have a good semicircular arc of flight. Feet kept low to the ground cause short strides and stumbling. Joints should flex freely — particularly the knees and hocks — although excess knee flexion can be undesirable, except in a carriage horse. The hind feet should be placed well forward under the body during each stride, particularly at faster paces. This is important, to give a good forward thrust. Movement at the canter should be analysed on both reins (different forelegs leading) by cantering in a circle: this demonstrated clearly how the horse places its feet and lifts them up. Working in a circle also gives a good indication of how the horse uses its back — whether it flexes and bends or is kept stiff. The tighter circle of a lunging ring may make an assessment of the suppleness of the horse's back easier. It may also be helpful to see the horse turned in a tight circle (in both directions) around someone standing at its shoulder: this checks its co-ordination and whether it can cross and use it hind legs properly. It is also work checking that the horse can back. While the animal is being put through its paces, ensure that these are true paces — the walk four-time, the trot two-time and the canter three-time (see Chapter 1).

FINDING THE RIGHT HORSE

Buying a horse that suits your requirements is as important as buying one that is healthy.

SHOULD I LOOK FOR SIGNS OF INJURY OR LAMENESS?

If you know what to look for, by all means run your hand over the horse's legs and feet. If you do not know what you are feeling, however, there is little point in doing this for it will be very obvious to the person selling the animal. Detecting problems and assessing their significance is really a vet's job, and is better left to one. A vet sees and treats such problems every day, and is generally much better able to evaluate them than you are. However, some horses can quickly be ruled out as unsound. It is always worth running a hand down the flexor tendons of the forelegs, looking for signs of heat or swelling indicative of a sprain. The foreleg fetlock joints are common sites of injury and arthritis, and can be checked for heat, pain on flexion, or a rounded appearance and swelling (articular wind galls). Swellings in or around the hock joint should be noted and viewed with suspicion. All four feet should be inspected, looking for signs of uneven wear of shoe or foot, as well as for normal foot conformation.

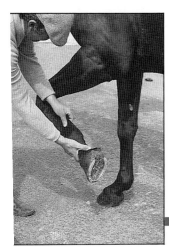

Feeling the legs for heat, swelling, and soreness *(left)*. All of these are signs of previous or current injury.

Spavin test By holding up the leg and flexing the hock joint for one minute before trotting the horse away, lameness resulting from a bone spavin may be more apparent *(right)*.

IS THE HORSE SUITABLE?

Before setting out, have a realistic idea both of what the horse is to be used for, and the type of animal best suited for the purpose. Having decided that you are happy with a horse's conformation, action and temperament, and that you like the animal, ask yourself whether it is really suitable (both for your intended use and for where you propose to keep it). Stick to your original specifications, and be prepared to pay a little extra for what you want rather than buy something unsuitable. If you go to buy a hack, don't come back with a racehorse!

WHAT QUESTIONS SHOULD I ASK THE SELLER?

When inspecting a horse for sale, it may not always be possible to fully assess its behaviour and temperament. It is therefore always advisable to ask a number of questions — whether the horse is easy to catch, for instance; whether it ties up and is quiet to shoe or clip; whether it loads in a horse-box or trailer; whether it has any vices. Information on its breeding, reputed age, and whether or not it has been officially measured and possesses a height certificate can also be useful. It is also useful to know whether or not the animal has been vaccinated, and if so, what against, when, and whether a certificate is available. In addition it is important to know about the animal's present feeding, so that there is not too sudden a change, should it move to a new home.

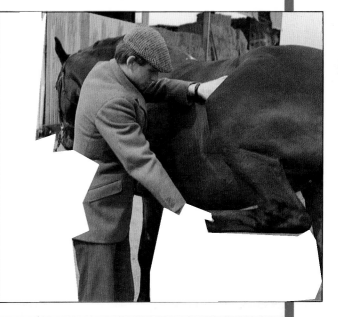

Vetting a horse *(right)*. Here the vet is **inspecting the horse's third eyelid** – a common site of particularly nasty cancerous growths. He is also checking the eye mucous membranes to ensure they are a healthy salmon pink colour. Paleness here indicates anaemia, yellowness – jaundice, and a blue tinge – shock.

❓ WHAT CAN I TELL FROM A HORSE'S TACK AND SHOES?

Pay particular attention to the tack used by the present owner or the person showing you the horse — if the animal is not being ridden in a simple snaffle bridle, for instance, why should a more severe bit be necessary? Take note of additional tack which may be used to help control the horse — why has it got a dropped, or grackle, noseband, or standing martingale? Does it have a breastplate, or crupper, in order to prevent the saddle slipping, or, if it has a numnah, why is that necessary? Look carefully at the horse's shoes: are they worn evenly and of a normal type? Excess wear at the toe of the front shoes may be a sign of navicular disease. The same finding in the hind feet is a sign that the animal does not pick its hind feet up well, possibly as a result of a problem in the hock joint. Special shoes are an indication that some defect in foot-flight is being corrected, and three-quarter shoes (for brushing) and rolled-toed hind shoes (for overreach) should be viewed with suspicion. Very broad, webbed shoes are often fitted to horses with chronic laminitis; horses suffering from this condition usually have excessive wear at the heels of the shoe (also sometimes found with ringbone). Protective hoof-pads and wedge heels should also be treated with caution.

❓ WHAT ELSE SHOULD I NOTE?

Do not be overimpressed with the performance put on by the person showing you the horse — the horse is presumably in familiar surroundings and doing what the person knows it does best, and it is unlikely that he or she will show you something it does not like doing. When you ride the horse, try not to repeat what the other rider has done — do something different. Ride it past the stable, to see if it is reluctant to go away from home (a sign of 'nappiness'); also, if possible, try it in traffic. Always be suspicious — check what feed the animal is having. A hot-tempered animal may become docile on hay and water, only to reveal its true form when you get it home and feed it concentrates. Likewise, if the horse has been ridden just before you arrive, ask yourself whether this has been done to calm it down — or to warm it up — to hide a problem. When arranging an inspection, it is best to ask to see the horse in the stable first, and not tacked up or just exercised. Allergy to mould in straw and hay causes a common and troublesome lung condition (COPD). If the horse has this problem it may need special bedding and food, and may cough or have serious breathing difficulties. Such a problem does not show at grass. If the horse looks fit and is being ridden from grass, ask yourself why; beespecially suspicious if it gives a harsh dry cough when entering the stable.

For further information about medical terms in this section, see also Chapter 8.

VETTING

Your vet is the best person to tell you whether a horse has health or injury problems. After his examination he will record any defects found and say whether they are likely to cause trouble in the future.

The heart and lungs are examined at rest during this stage. The animal is then taken out and its conformation studied. Its action while being led in hand is analysed at the walk and at the trot, and when turned in a tight circle or made to back. It is then exercised — normally by being ridden, although unbroken horses are occasionally lunged. Here, the action at the canter and at the gallop are assessed. The exercise must be fairly strenuous so that the wind can be listened to, and the heart and lungs checked during exertion. If the animal is to be used for jumping, some jumps may be included in the examination. Finally, the horse is returned to its stable and rested for 15 minutes. During this stage, its sight and hearing may be checked. It is then trotted up in hand once again to ensure that no lameness or stiffness has resulted from the exercise.

WHAT DOES VETTING ENTAIL?

A vetting is done on behalf of the buyer never for the seller. The horse is always examined from the point of view of its suitability for a specified purpose, so it is important to tell the vet what you intend to use the horse for when you request the examination. Also let the vet know if there is anything that particularly concerns you about the horse. A standard examination procedure is used; the fee is based on the time taken and not on the value of the horse. Your vet should be able to give you an exact estimate of the cost beforehand.

The horse is first examined in the stable. Each part of the animal is inspected thoroughly, examined, felt, and any abnormality is recorded.

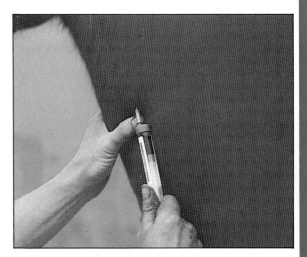

The vet takes a **blood sample** from a vein in the horse's neck. As with humans, this could be to test any number of disorders.

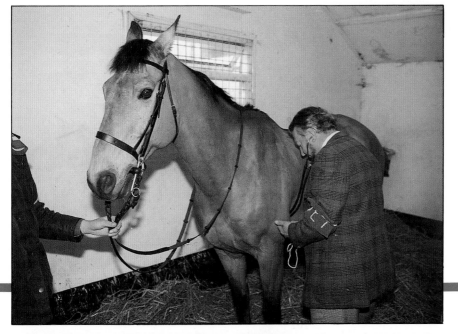

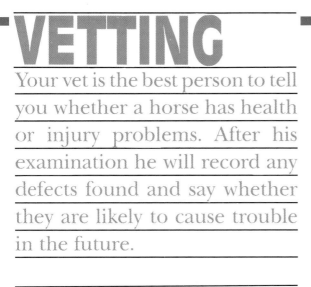

Vetting a horse. Your vet is the best person to tell you whether a horse has a current health or injury problem, or a previous one that is likely to recur *(left)*. Following a thorough examination, the vet will record any defects found, and give an opinion whether they are likely to cause trouble in the future, and whether the horse is suitable for your intended purpose.

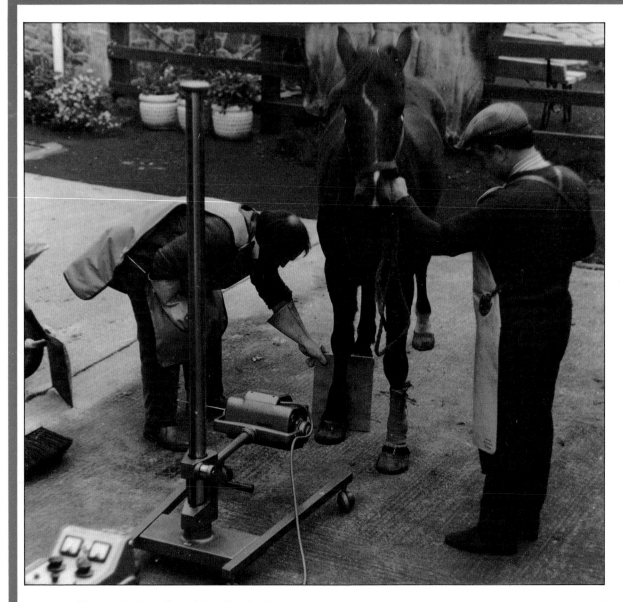

A certificate is issued to identify the horse, noting every abnormality or conformation fault detected, and expressing an opinion on their significance. Finally, the vet states whether or not, in his opinion, the horse is suitable for the purpose for which it has been examined — a few minor blemishes are of little importance for a hunter or hack, but may make an animal totally useless for showing.

Occasionally, there may still be doubt about an animal's soundness or suitability after vetting, and extra tests might be required. These could include examinations by X-rays, electrocardiography or endoscopy of the bronchial airways. For especially valuable animals, X-ray examination of joints and feet may be considered.

? SHOULD I HAVE MY HORSE EXAMINED BY A VET BEFORE BUYING IT?

Even if you really like a horse, there is little point in buying the animal if it is going to be lame, ill, or totally unsuitable for your requirements. Although an experienced horseman can certainly spot most major defects, it is a vet's job to evaluate and deal with disease, lameness and injury. He is much more used to picking up early signs of

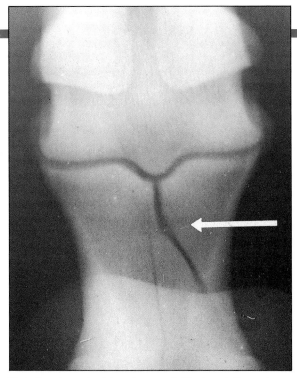

X-raying a horse that is lame in its off-fore (left).

The results of the X-ray (above), showing a clear **fracture of the pastern**. Whereas once this would probably have meant the end of the horse's life, modern surgical techniques have enabled this fracture to be pinned successfully, and the horse was able to resume a normal working life.

on the age and value of the horse; if such an examination is required, it is not a substitute for a pre-purchase vetting — it is designed only to ensure that the animal is healthy. Loss-of-use insurance does require a full veterinary examination: the vet should be informed if you intend to take out this type of insurance before he or she carries out the pre-purchase examination so as to avoid the cost of two examinations. Veterinary fees can be expensive, particularly if the animal requires surgery. Policies to cover this eventuality are worth while considering. There is generally an excess clause (a normal rate is £50), so read the small print carefully before taking out the policy. It may be difficult to insure horses or ponies aged over 12 years unless you have already had a policy with the company for some years.

? WHO IS LIABLE IF YOUR HORSE INJURES SOMEONE OR DAMAGES PROPERTY?

The law is complicated in relation to horses and any damage they cause. If your horse strays from its paddock into someone else's property and causes damage, the owner of the other property can sue you for damages, whether or not you have been negligent. If your horse gets loose on the road, or kicks and injures someone, those injured in property or body may have to prove negligence on your part before they can be awarded damages. Leaving a gate open, or the stable door unlocked, or riding a horse known to kick through a crowd could all constitute potential forms of negligence. There is an increasing tendency for people to take out insurance also to cover legal fees, and, as a result, readily to sue for damages. All horse-owners are strongly advised to cover themselves against such claims by taking out Third-Party Public Liability insurance. This cover is often included with other horse insurance policies — mortality, loss of use and so on — and is incorporated as an additional benefit of membership of the British Horse Society. Many — but not all — householder's insurance policies include Third Party Public Liability cover, which also includes an owner's horses.

As with all insurance policies, you should consult a broker or read the small print carefully to see how you are covered. In the case of an accident, an insurance company pays out only if it is legally obliged to do so. It is not obliged to refund any damages you may have paid through a sense of moral obligation following an accident .

trouble, and, from his experience in practice, of knowing what is likely to cause trouble and what is not. A veterinary pre-purchase examination ('vetting') is the equivalent of a full structural survey before buying a house. It is designed to examine the horse in detail as thoroughly as is feasible in a given time, and to find as many defects as possible. It cannot detect every conceivable fault, but is intended to rule out as many as possible. If you have decided that you like everything about a horse, get your vet to examine it before concluding the deal. His fee could save you a lot of trouble later.

? SHOULD I INSURE MY HORSE?

It is very advisable to insure your horse from the moment at which you conclude the deal — before the animal is transported to your premises. You should consult an insurance broker for details, but in essence you can insure your horse against death (or destruction on humanitarian grounds), for loss of use, and/or for veterinary fees. Mortality insurance may or may not require a veterinary examination, depending

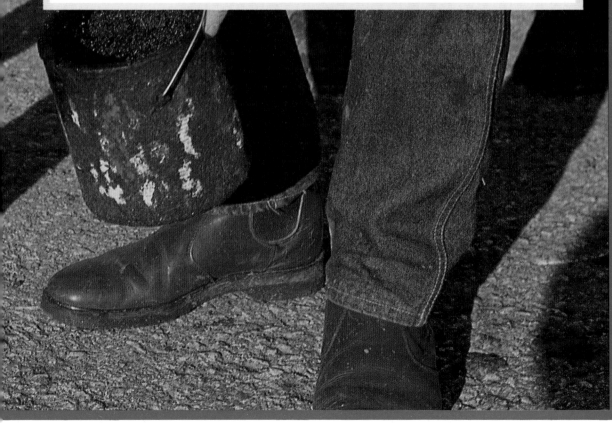

SECTION 2
SAFEGUARDING YOUR HORSE'S HEALTH

EMERGENCY MEDICAL CARE

7

Spotting when a horse is ill or injured, and recognizing when to call the vet, can be difficult. Knowing what to look for, and how to carry out simple procedures like taking a horse's temperature, can lessen the worry of this responsibility. Horses are also very liable to suffer kicks, knocks, superficial wounds and other minor injuries which the owner has to deal with. Having the right equipment to hand, and using it correctly, is essential. Knowing how to restrain an injured horse, and what to do in an emergency until help arrives, could even save an animal's life. The job of administering wormers and some medicines also usually falls to the horse's owner, as does the task of nursing it during convalescence. It is important to know how to carry out these tasks. Doing so incorrectly could make matters worse.

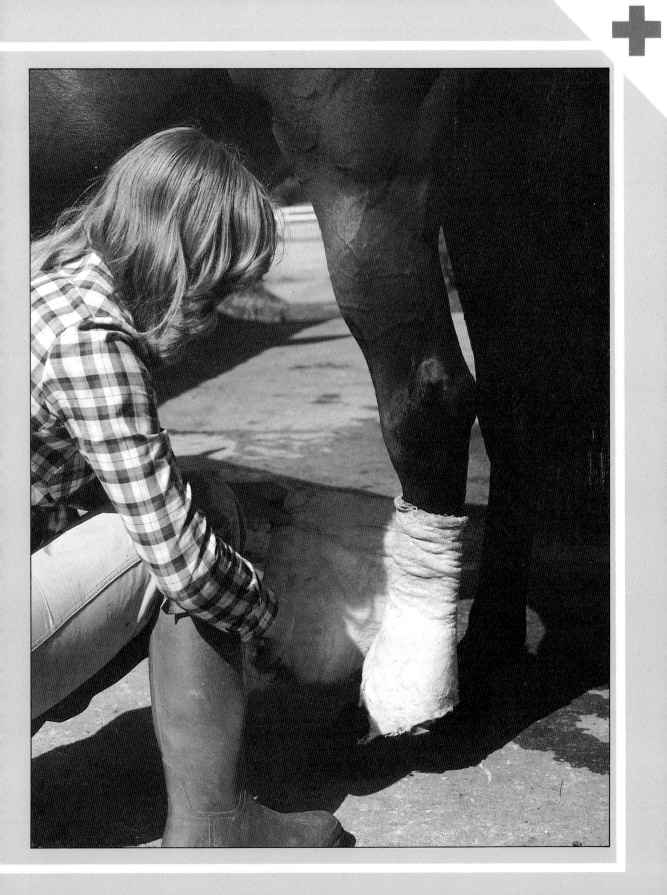

SIGNS OF ILLNESS

Changes in the horse's condition or behaviour may be the first signs of ill health. If an illness can be detected early, later developments are often less serious.

? WHAT SIGNS OF ILL-HEALTH MAY BECOME APPARENT IN MY HORSE?

Signs of ill-health may be detected by changes in an animal's appearance — in its body condition or coat, or any abnormal enlargement or swelling. Changes in behaviour — including loss of appetite, dullness or nervous signs — are also indicative that something may be wrong. In addition, alterations in bodily functions — manifested perhaps in abnormal discharges, difficulties in breathing, or abnormal passing of urine or droppings — should be viewed with suspicion. Reluctance to move, or changes in gait, may indicate ill-health or lameness.

Horses are creatures of habit. Careful observation when checking the animal each day is important to spot any small variation from its normal behaviour that may show that something is amiss. If an illness can be detected in its very early stages, later developments are often much less serious.

A horse's coat is a good indicator of its state of health — the gloss being lost when the animal is ill. Patchy sweating is always a sign that something is wrong. Dehydration affects the skin; a useful test for dehydration uses the fact that when the skin of a dehydrated horse is pinched it is very slow to return to its normal shape.

A generalized loss of body condition is often a sign of chronic illness. In some conditions (such as chronic liver damage from ragwort poisoning), this may be the only sign. In other conditions, changes in the shape of the body may be seen — for example, a 'pot belly' is often the sign of a serious worm infestation. Accumulations of lymphatic

fluid beneath the skin may cause swelling of the legs or abdomen, and are a sign of poor circulation, insufficient exercise or infection.

The appetite is perhaps the best guide to health. If a horse fails to eat up properly, it is always worth trying to find a reason.

Stiffness and reluctance to move can be a sign of disease (such as tetanus or laminitis) as well as of injury or lameness. Resting a foreleg is also abnormal. Dullness, lassitude and general lack of interest in what is going on are also behaviour changes that need investigating. More obvious nervous signs — such as tremors, paralysis, lack of co-ordination or head pressing — indicate something serious, and that professional help is needed.

Change in body functions such as in breathing — when respiration becomes rapid, laboured or noisy, with or without coughing and nasal discharges — are also symptoms of illness. Excess drinking, too frequent or infrequent urination, loose or hard droppings (or an absence of droppings), and drooling saliva or dropping food from the mouth ('quidding'), are all also signs for concern.

SKIN PROBLEMS

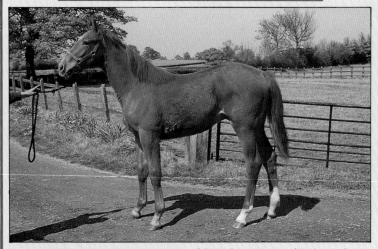

A horse's coat is one of the best indicators of its state of health. General lack of condition and neglect of grooming has resulted in the poor, dull-looking coat of this horse (*above*).
Insufficient feeding (*left*) has led to **malnutrition** in this horse, and it is also showing signs of skin disease.

? MY HORSE IS BECOMING THIN: SHOULD I WORRY?

Keeping condition on a horse is often quite a problem, particularly in fine-bred individuals who are finicky feeders. The first consideration is obviously whether the animal is receiving enough food. If it is having regular, strenuous exercise, weight loss may be due to the fact that the animal is receiving insufficient concentrates in its ration. It is often very difficult to keep condition on Thoroughbreds outdoors in winter: they can become 'poor' in spite of receiving large amounts of extra food. This is because their thin skins provide little protection from the elements, and the extra energy intake, and more, is utilized in simply maintaining the body temperature.

Worms are a primary cause of unthriftiness (failure to thrive); their presence can be detected by tests. Regular worm treatment of all horses, especially those at grass, is essential to keep them healthy.

Digestive disorders can also be associated with weight loss. This problem can occur with chronic damage to the lining of the intestine, which affects the absorption of nutrients. Likewise, chronic diarrhoea and chronic liver damage may also cause horses to become thin. Dental problems sometimes result in inefficient grinding of food and poor digestion.

? WHAT STEPS SHOULD I TAKE TO REMEDY POOR CONDITION?

If the animal appears to be losing weight, first check its diet, and try giving it extra food (in the form of concentrates). At the same time, give it a worm dose of a different type from the one last used — worms sometimes develop drug resistance. If no improvement occurs soon (that is, within two weeks) consult a vet. He can check on the horse's teeth and, if necessary, treat them. He may also take samples for testing, to tell whether or not something is wrong. Blood tests are particularly useful in this respect: they can indicate not only liver or bowel damage, but also overwork, anaemia, and mild bacterial or viral infections that may be showing few other symptoms. Loss of condition is a sign of many chronic diseases. If extra feed and worming do not produce the desired results, get the horse checked sooner rather than later. Delay in detection and treatment of some chronic diseases can make them very much more difficult to cure, and could permanently affect an animal's subsequent health and performance.

? MY HORSE'S COAT HAS LOST ITS SHINE: WHAT SHOULD I DO?

Except when at grass in winter, a horse's coat should always have a glossy sheen. A dull coat usually means that something is wrong, and may be a sign of a serious worm infestation or an internal disorder such as chronic liver or kidney damage. Horses with a fever rapidly lose shine from their coats. Some mineral and vitamin deficiencies are also reflected in a poor coat. To counter the latter condition, a few commercial mineral and vitamin supplements are available which are specifically designed to improve horses' coats; many of them are indeed very effective in restoring a glossy, healthy coat. Washing a horse is not recommended unless it is necessary for some medical reason; it removes the natural oils, and their waterproofing benefits, from the coat.

? WHY IS MY HORSE OFF ITS FOOD?

Loss of appetite can be one of the first signs of illness, so if a horse leaves its food and appears dull or 'off-colour' it is sensible to check its temperature. If the horse is interested in food, but fails to eat it or drops it from its mouth ('quidding'), chokes or brings it back up through its nose, a vet should be called to check on its teeth and mouth. When a horse fails to eat, it is also advisable to look at its droppings. If the droppings are very hard and the quantity is small, the animal is probably suffering from an impaction (constipation). This is a

common problem in stabled horses, and bran mashes and other laxatives (mild 'physics' such as Epsom or Glauber's salts) are sometimes given to prevent this.

Horses have highly developed senses of taste and smell. They do not like stale food and may not eat fresh food when mixed with it. Mangers should be cleaned out before each feed, and all uneaten food be removed before fresh food is added. Fussy feeders can sometimes be tempted by adding sweeteners such as molasses to the ration. If a horse eats nothing at all for 24 hours there may be something seriously wrong, and you should consult a veterinary surgeon immediately.

? CAN I TAKE MY HORSE'S TEMPERATURE?

All horse-owners should be able to take their animals' temperature. It can be done easily, using a normal clinical thermometer, and is by far the best way of telling when a horse has an infection or may be developing a fever. A thermometer is thus a basic requirement for a first-aid kit or tack-room medicine cupboard.

Before proceeding, the thermometer must be shaken vigorously to ensure that the mercury level starts below 37.4°C (100°F). It is then lubricated (with petroleum jelly or saliva), inserted into the horse's rectum and held there for one minute, before the temperature can be read. To carry this operation out, get someone else to hold the animal's head firmly (preferably using a bridle), and stand beside and just in front of its near (left) hindquarters. Moving quietly and talking to the horse so as not to frighten it, run your hand along the quarters to the root of the tail. With the left hand pull the tail towards you, exposing the anus, and insert the thermometer with the right hand. The tail can be released at this stage, while the thermometer is held in the rectum. To stop the horse moving away sideways and pulling the thermometer out, however, it may be helpful to keep hold of the tail, or to hold the horse against a wall (on its right side). With nervous horses, or those that try to kick, it is helpful if the person who holds the animal's head also holds its near (left) foreleg up. In any case the thermometer must be held all the time it is in the rectum; if it is not, its presence may stimulate the anal sphincter either so that droppings are passed and the thermometer is smashed on the ground, or so that the anus relaxes and the thermometer disappears inside!

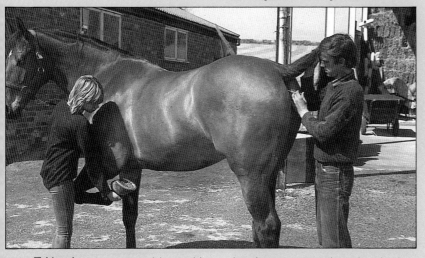

Taking the temperature. It is essential for a horse owner to be able to take a horse's temperature to determine when it has a fever (above).

Taking a horse's pulse by feeling the maxillary artery where it crosses the jaw (*below*). A raised resting pulse can be a sign of shock, but interpreting the different types of pulse associated with circulatory problems is a job for the vet.

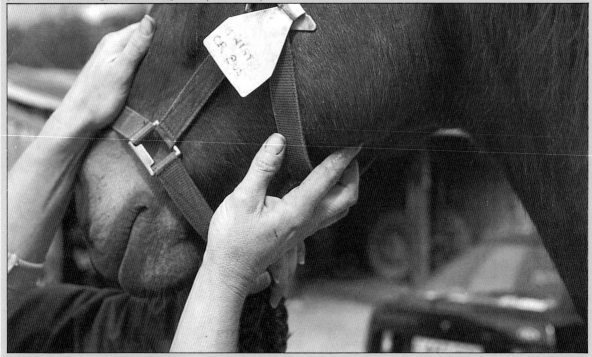

 WHAT SHOULD THE NORMAL TEMPERATURE BE?

The normal temperature of a horse is 38°C (100.5°F). The temperature can vary slightly up or down by half a degree during the day, and between individuals. Foals have a slightly higher temperature than adult horses; for them up to 38.6°C (101.5°F) may be considered normal. Taking the temperature is often the only way of telling that a horse is ill, particularly when it is incubating an infection. Fevers and subnormal temperatures (as from shock) require prompt veterinary attention. A vet should be called if a horse has a temperature above 38.6°C (101.5°F), or below 37.7°C (100°F).

WHAT IS THE NORMAL BREATHING-RATE FOR HORSES?

At rest, a horse normally breathes about eight to 12 times per minute. This resting rate is greatly increased in pneumonia or pleurisy. Injuries to the ribs (sometimes seen in foals whose ribs are damaged at birth) may alternatively cause rapid breathing; and the rate of respiration is also increased if the horse has a fever. When an animal is excited, there is little point in attempting to measure its breathing.

In allergic respiratory disease(COPD), the character of breathing itself alters, and the animal has to use an extra expiratory effort to empty its lungs. Normally, expiration is a passive process, involving relaxation of the ribcage. In COPD, an extra contraction of the chest muscles (forced expiration) follows normal expiration. The horse is said to have a 'double-lift' of the ribcage which can best be seen from the front and slightly to the side (standing by the near shoulder).

WHEN SHOULD I CALL A VET?

It is always difficult to know when to seek professional help. As a rule, it is better to do this sooner rather than later, for delay, in many conditions, may have serious consequences. Where there is any doubt, it is best to contact the vet, giving him as many details as possible (whether the horse is off its food, has a temperature, is sweating up, or is behaving abnormally), discussing with him whether or not a visit is necessary and what you should do until he arrives. For nearly all cases of colic the vet should be called immediately. The sooner painkillers are administered, the less chance there is of complications such as a 'twisted gut'. Horses with a fever — a temperature of more than 38.6°C or (101.5°F) — also need veterinary attention, as do those suffering from acute lameness or injury. To have a chance of healing successfully following stitching, wounds must be treated when fresh. Accordingly, it is important to call a vet as soon as a cut is discovered: there is little point in asking him to stitch it on the following day.

MY HORSE MAY BE OFF-COLOUR OR ILL: WHAT SHOULD I DO?

If, when visiting the paddock or stable to check it, your horse appears unwell (see *Signs of illness*, page), it is important to spend some time observing its behaviour. If the animal is at grass, it is best to move it into a stable to do this. Watch its breathing; if possible take its temperature; offer it some fresh food; then, if it is stabled, look for droppings and note their consistency. Shake up the bedding and see if the animal stales (urinates). You should then contact a vet; tell him what you have observed and ask him to call.

NURSING

Keeping a sick horse warm, well fed and comfortable is the owner's task. When an animal is off colour, a special diet may be needed; whilst rugs and bandages will provide warmth and protection.

? MY HORSE IS SICK: HOW SHOULD I LOOK AFTER IT?

Horses that have a fever, or are suffering from an infectious disease, require special nursing. Loss of appetite is often one of the first signs of a fever, and it is important to keep sick horses eating. All old, or uneaten, food should be removed from the manger regularly, and small amounts of fresh food used to tempt it to eat. Where possible, this should be fresh, green food — cabbage, chopped apple, picked grass (not mowings), carrots — to try to keep the horse interested.

Another complication of a fever is that bowel movement slows down, and there is a tendency towards impaction (constipation). Green food should help to overcome this problem, but hot bran or linseed mashes are also beneficial for their laxative effect. Oats should not be fed to sick horses, as they are too rich for an animal that is not getting any exercise when ill. Compound nuts or cubes may be fed, to retain a good body condition. Adding molasses to the concentrates may tempt a sick horse to eat. When an animal has a temperature, it is important to keep it warm. Usually, a light rug is sufficient, and the legs can be bandaged (woollen stable bandages are best) for warmth.

A good deep bed is essential. If straw is used, it can be banked round the edge of the box to reduce any draughts, and may help prevent the animal becoming cast (stuck when lying against a wall).

? HOW DO I PUT ON A PROTECTIVE BANDAGE?

Bandaging is used to protect the legs from injury (as exercise bandages or travelling bandages); to keep the animal warm when ill; to treat and protect leg

LEGS

Bandaging a horse's legs to protect them in the stable or during travelling. A layer of gamgee is wrapped around the leg and a stable bandage wound down over it,

The end of the bandage is turned down, and the bandage wound over it to secure the end. Bandaging then continues on down the leg.

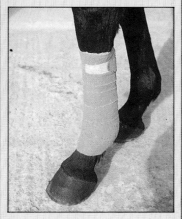

The bandage is taken down over the fetlock to the coronet, and then wound back up the leg. The end of the bandage must be secured on the outside of the leg.

wounds; and to support sprained tendons. They are rolled with the tapes, if any, on the inside before use. Bandages are commonly put on over cotton wool or, even more often, over gamgee tissue (cotton wool between gauze). This is placed round the leg first and the bandage applied (beginning just below the knee or hock), in the same direction as the overlap of the gamgee (see photo/diagram). The bandage must be fairly firm but not too tight. It is wound around the leg at an oblique angle, down to the fetlock for exercise bandages, and to the coronet for stable and travelling. The process is continued up the leg, to end below the knee or hock, where the tapes are tied on the outside of the leg or the end is fixed with a safety pin. Exercise bandages can be taped with Elastoplast, or stitched, to ensure that they do not become loose during work — but be careful: there is a danger that exercise bandages may actually cause tendon injury, rather than prevent it, if put on too tight. Cotton wool can sometimes 'ball up' and exert uneven pressure on underlying tissues. For this reason, pieces of foam sponge are preferred under exercise bandages.

? HOW SHOULD I BANDAGE A WOUND?

Bandaging is often an integral part of wound treatment. For injuries below the hock or knee, this presents no problems; but for injuries higher up the leg, it is often difficult, or even impossible, to keep a bandage in place without causing pressure problems. If the bandage is tight enough to stay on the leg, it is actually *too* tight and likely to cause pressure sores. Thus, on the hind limb, it is not practical to bandage wounds above the hock. On the foreleg, it is impossible to bandage round the elbow region. However, wounds on the forearm and on the front of the knee — a common site of injury — can be bandaged. An Elastoplast bandage may be required around the upper part of the dressing to stick it to the skin and keep it in place. It is vitally important *never* to bandage over the bone at the back of a horse's knee joint (the accessory carpal bone): to do so almost inevitably causes skin damage in the form of pressure sores resulting from injury when the animal bends its knee getting up and getting down. To protect a wound on the front of the knee, the front of the joint is thus covered in bandage, and the back is left unbandaged but covered merely in the dressing.

WHY IS IT NECESSARY TO BANDAGE BOTH A BAD LEG AND A GOOD ONE?

The horse's relatively top-heavy muscular body, with its thin lower limbs moving on a single toe, is ideally designed for speed. However, this arrangement has a distinct disadvantage if one limb (particularly a forelimb) is injured, and the horse has to rest it. The total weight of the corresponding upper part of the body must then be born on the opposite limb, whose thin extremities are not designed to withstand this extra loading for any length of time. This in itself can cause damage, or damage may happen as a result of the extra strain when the animal tries to 'save' itself when getting up and down. It is thus essential to put an additional supporting bandage on the leg opposite the one which is injured, especially with foreleg injuries such as sprained flexor tendons. If the horse is unable to move much, supporting bandages on all four legs may help to improve the overall circulation and thus prevent them swelling up with lymph fluid. In the case of serious injuries which require prolonged rest — such as a limb fracture — strong support for the 'other' leg is very important. Regrettably, there have been many instances in which it has been necessary to destroy horses because of damage to the 'good' leg (where, for example, the pedal bone has been pushed through the sole) even though the original fracture was healing well.

HOW DO I PUT A TAIL BANDAGE ON MY HORSE?

Tail bandages are normally 64–75mm in width (2.3–3 inches) and made of crêpe or stockinette; they are used to improve a horse's looks and also to protect the tail when travelling. Serious damage can result if tail bandages are put on too tight or left on too long. They should *never* be left on overnight.

The tail is usually dampened, using a water brush, and a short length of the rolled bandage is unrolled. The end is placed under the upper part of the tail, and two turns are used to secure it. A further two turns are made above this to gather in the hairs at the root of the tail, and the remainder is applied down the tail, finishing at the end of the tailbone, where the bandage is tied. To remove it, the tail bandage is grasped with both hands at the top and slid down the tail.

BANDAGING

The elasticated bandage is unwound, and a short length held out over the horse's rump *(left)*.

The bandage is wound once around the horse's tail *(right)*, with the end of the bandage still being held in the left hand.

The end of the bandage is turned down over the bandaged tail *(left)*, and another turn of the bandage is taken over the top of it.

The bandage is wound down the tail to the bottom of the dock *(right)*. Any surplus bandage should be wound back up the tail.

The tapes should be tied neatly on the outside of the tail *(left)*, with the ends of the tapes tucked in.

The bandaged tail should be gently bent into a comfortable position *(right)*. The bandage is removed by taking hold of it at the top of the tail and sliding it down.

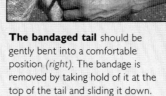

MEDICINE

The vet will prescribe the appropriate medicine for a sick horse, but it is often the owner's job to administer it. This is not always an easy task, since horses often refuse to eat 'doctored' food. Using the right techniques will help your horse to recovery.

? THE VET HAS PRESCRIBED POWDERS FOR MY HORSE: HOW DO I GIVE THEM?

Horses have very well-developed senses of taste and smell. They have an incredible knack of being able to detect 'doctored' food and thereafter refusing to eat it. When giving worm powders or other medications (such as phenylbutazone – 'bute') that have been prescribed, it is essential to try to disguise the taste. A mash, of bran or linseed, is a very good way of doing this. The sweet taste of molasses is also an excellent way to disguise medicines, especially those with a bitter taste. Horses are quite partial to bread, and enclosing a powder in a sandwich is another useful trick.

? HOW DO I GIVE MY HORSE A WORM PASTE?

Worm pastes, in a pre-packed dispensing tube, are an ideal way to treat horses for worms. However, there is nothing more annoying than to see an expensive dose of wormer being spat on the floor! It is therefore worth getting some help and taking a little time and trouble to do it properly. To get the horse to swallow the paste, the medication must be placed on the back of the animal's tongue. For a right-handed person it is much easier to administer the paste to the right ('off') side of the horse's mouth. If someone else is holding the horse on the near side (with a head collar and leading rein – not a bridle because the bit gets in the way), stand on the off side, place your left hand on the side of the horse's face, with fingers on the bridge of the nose, and your thumb inside the horse's mouth in the space behind the incisor teeth and pressing against the roof of the mouth. This ensures that the horse opens its mouth. With the right hand insert the tube, pushing the end to the back of the mouth, and press the plunger quickly. Withdraw the syringe and, placing the right hand under the jaw, hold the horse's head up until it swallows.

? MY HORSE IS COUGHING: IS THERE ANYTHING THAT I CAN GIVE IT THAT WILL HELP?

Cough medicines and electuaries can be helpful to soothe the throat of coughing horses. Cough linctuses are best administered using a syringe placed in the side of the mouth and squirted to the back of the tongue. An old worm paste syringe or a 20cc disposable plastic syringe is ideal for the purpose.

An electuary is a thick paste with a glycerine or black treacle base, containing one or more compounds (such as potassium chlorate, friar's balsam, camphor) which can soothe a horse's throat. It must be smoothed on to the back of the tongue. A wooden spatula or the back of a tablespoon is suitable, although it may be necessary to catch hold of the tongue and pull it out through the side of the mouth in order to do this.

It is always important to know why a horse is coughing. If the cough is due to an allergy (COPD), cough mixtures are a waste of time. If a coughing horse has a fever or a thick nasal discharge, antibiotic treatment by a vet is necessary.

? SHOULD I GIVE MY HORSE A COLIC DRINK, OR DRENCH?

In the past, various patent remedies for colic were popular. Most of these were fairly ineffective and were administered as drenches, using a long-necked wine bottle, and holding the horse's head up in the air. There is some danger in doing this – the horse may choke and inhale some of the liquid into its lungs. This can cause 'inhalation' pneumonia which, if a large amount of fluid enters the lungs, can be fatal. Drenching horses should be attempted only by experienced horsemen – and even then there is some danger, particularly if the animal is already distressed (through colic perhaps). If liquids are required, they are usually needed in such large volumes as are best administered by a vet using a stomach tube and pump.

Giving a horse a worm dose is a routine task for horse owners *(below)*.

OTHER TREATMENT

Heat treatment or hosing down with cold water are often the best ways to help heal injured tissue. Deeper tissue damage may require more specialized forms of treatment.

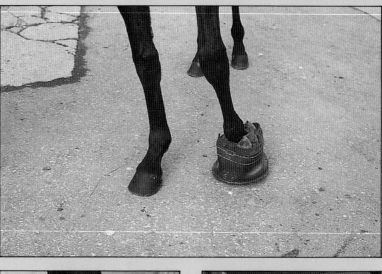

? WHAT IS HOT FOMENTATION?

Hot fomentation is the process of applying heat to the skin to increase the blood supply to the underlying tissues. This is done either to reduce an inflammatory swelling, or to try and bring an abscess to a head. Hot fomentation is applied by using hot water directly (by 'tubbing' — standing a limb in a bucket), or by using it indirectly to heat a cloth or poultice which is then put on the affected area. Fomentation is most useful for bringing an abscess to a head, or for drawing poison from an infected wound.

? WHAT IS A POULTICE AND WHEN SHOULD I USE ONE?

A poultice is a convenient way of applying hot fomentation. Some forms of poultice make use of additional chemicals which help to draw fluids from the body — for wounds and sprains, for example, a kaolin poultice is frequently used to help draw fluid from the affected area. Kaolin (china clay — hydrated aluminium silicate) comes as a thick, clayey paste which must be heated to make it soft before application. This is usually done by immersing a tin of kaolin in boiling water. The kaolin is applied around the affected area, using a spatula or tablespoon, after which a waterproof layer (a plastic film) is placed around it, held in place with a bandage. Kaolin is also used for reducing the inflammatory swelling of sprains. Poultices may also be used to soften a horse's foot or to draw out pus from the foot, which is a common cause of lameness. In the past, a bran poultice was sometimes used: boiling water was added to bran, which was then put in a canvas (or plastic) bag and

placed round the horse's hoof, in turn then being inserted either into a large poultice boot or in some sacking and then bandaged round the fetlock to hold it in place. A specially impregnated cotton-wool pad ('Animalintex') is now more often used for this purpose, soaked in hot water (to blood heat) and applied to the hoof under a plastic bag and leather poultice boot. Alternatively, Epsom salts and glycerine can be applied to a piece of gamgee and bandaged to the foot, and is very effective for drawing out poison.

Poulticing the feet make them soft. When pus in the foot is suspected, vets therefore sometimes request that the effected foot be poulticed for 24 hours before their visit. This makes it very much easier to cut the sole to locate the pus, and to make a clean incision through which it can drain.

? WHEN IS HOSING WITH COLD WATER HELPFUL?

Injury to joints, tendons, and ligaments, either from trauma or the repeated affects of concussion (wear and tear), results in inflammatory changes in these structures. This produces an inflammatory fluid exudate which causes swelling, pain and heat in the affected area. In the early stages of injury, it is thus helpful to try and reduce this fluid exudate, and applying a continuous stream of cold water from a hose is a very effective way of doing this — indeed, for joint and tendon injuries, such treatment can be very effective in the early stages. The best results are achieved by hosing for twenty minutes at least twice a day. Another form of therapy — which is sometimes combined with hosing or used as an alternative — is ice-packs. These are specially-manufactured, re-useable plastic bags which contain a liquid that can be frozen, and which can be bandaged round a horse's leg. They are often used for the initial treatment of sprained flexor tendons, where it is important quickly to reduce any initial reaction caused by haemorrhage within the tendon.

A hot poultice is held in place on the foot with a poultice boot *(left)*. The poultice softens the foot, and then draws pus from an infected sole abscess.

Hot fomentation. A poultice is applied to a wound on the fetlock *(far bottom left)*. The poultice is soaked in hot water before it is applied to the leg. It is held in place by a bandage.

Hosing with cold water is done to treat a sprained foreleg flexor tendon injury *(bottom left)*.

Hot tubbing. Here the foot is held in warm water for twenty minutes, to help draw poison from an infected injury *(right)*.

 WHAT OTHER FORMS OF TREATMENT MIGHT HELP MY HORSE RECOVER FROM INJURY?

The important role that physiotherapy often plays in recovery from human injury has led, increasingly, to a realization of similar applications to equine injuries. Although poultices, cold hosing and ice-packs are effective for the more superficial injuries, they are unable to have much effect on deeper damage. Ultra-sound and short-wave radiation machines have been used for many years to help disperse blood clots, reduce swelling and soothe inflammatory reactions resulting from tissue damage. This form of treatment is ordinarily applied twice daily in 15-20 minute sessions in the early stages soon after injury, and can be particularly helpful with both joint and tendon injuries.

It has been known for some time that there are small differences in electrical potential between different parts of the body. It is now thought that these may have an important role in the repair and restructuring of tissues after injury. It has also been suggested that healing can be improved by the influence of a magnetic field on the electrical potentials

within the injured tissue. Devices are now available to apply a continuous or pulsed magnetic field to damaged tissue. The field can be produced by an arrangement of magnets or by passing an electric current through wire coils, and has been used as a form of treatment for a wide variety of horse-injury problems recently. The beneficial effect, in human medicine, of improving and accelerating the healing of bone fractures is well documented, and similar benefits are evident in horses. Such devices have also been widely used for many different types of soft-tissue injury (sprained tendons, ligaments, and so on), but there has been no scientific appraisal of the benefits of this usage as yet.

 CAN LASERS OR PHYSIOTHERAPY HELP HEALING?

The search for ways of penetrating deeper into tissues to reduce deep-seated inflammation has led to the use of lasers in equine veterinary practice and physiotherapy. These have been shown to have benefits for treating wounds and to

accelerate the healing process considerably. They have also been used to treat a variety of inflammatory problems (arthritis, sprained tendons and ligaments) and would appear also to achieve good results. Like electromagnetic therapy, this form of treatment is currently very much in vogue, but its usefulness for specific injuries in the long term remains yet to be evaluated.

It is important to have an accurate diagnosis of the cause of a problem before beginning physiotherapy. Once treatment has started, swellings or other outward symptoms may disappear, and it can be impossible to make a diagnosis or to give a meaningful prognosis (estimating the chance of recovery and the time it could take). The vet is the best person to give a diagnosis, and you should consult him first. There are qualified physiotherapists who specialize in horse injuries, and their services and experience are often most helpful. Owners should be wary of buying physiotherapy equipment when their horse is injured: in most instances suitable equipment for the more common problems can be hired.

? ARE FORMS OF ALTERNATIVE MEDICINE AVAILABLE FOR HORSES?

Like its human counterpart, veterinary medicine has always had a 'fringe' element. Herbal remedies, homeopathy, and acupuncture have all been used on horses, as has manipulation by osteopaths and chiropractors. There has been increased interest in alternative medicine for horses recently, coinciding with the similar development in human medicine. No proper evaluation of the effectiveness of the different forms of therapy has been undertaken, however, although many disorders have been successfully treated by all these means. Acupuncture is used to anaesthetise and treat horses in China but, so far, has found little favour in the West, except for a few vets who use it to treat some forms of lameness. As in human medicine, people often turn to alternative medicine for complaints where conventional medicine has had poor results. Back problems are one example of this, although it is highly questionable whether it is possible to 'manipulate' the bones in a horse's back — these are in many instances fused together. The benefit of such procedures may well lie in the relief from pain that follows the acupuncture-like effect of pressure exerted at the correct spot.

? CAN A 'BLISTER' ACCELERATE HEALING? IF SO, WHEN SHOULD ONE BE USED?

After the initial inflammatory reaction has died down (in two to three weeks), the rate at which an injury heals is dependent on the blood supply to the area. In an effort to increase the blood flow, a method of applying irritation to the skin above the area is sometimes used. This process is known as counter-irritation. In the past, this was a popular method of treating many horse injuries, and 'firing' with a hot iron and 'blistering' were used for the purpose. Although firing is now seldom used, blistering is sometimes helpful to accelerate healing and to reduce swelling in such conditions as splints, ringbone, 'curbs', or sprained flexor tendons.

Milder forms of blister are usually in liquid form, and are applied daily to the affected area. They are commonly known as 'working' blisters. These are sometimes also applied to the coronet to increase the rate of horn growth. Stronger blisters (red or green)

A bib muzzle. This is attached to the headcollar to prevent a horse getting at bandages or chewing its rugs *(above)*.

are used under veterinary supervision for treating splints, ringbone or sprained tendons. These blisters must be rubbed well into the affected area after it has been clipped. The horse must be prevented from getting at the area, or getting blister on its face. The reaction provoked by a blister depends on how long and how vigorously it is rubbed in, as well as its strength (concentration) and the thickness of the horse's skin.

? HOW CAN I STOP MY HORSE GETTING AT A BLISTERED AREA, CHEWING ITS RUGS, OR UNDOING BANDAGES?

Although a bib — a piece of leather or plastic attached to the three D-shaped attachments of the head-collar — may stop a horse chewing its rugs or undoing a bandage, it is not sufficient to prevent it from getting at an area that has been blistered. Likewise, even though a muzzle may stop an animal biting the skin with its teeth, it may still rub it with the muzzle. After blistering, it is usual to tie a horse up 'short' to the wall or manger for a while before loosing it and fitting a cradle. This is a series of wooden battens attached to two leather collars which go around the top and bottom of the horse's neck. The battens prevent the neck from being flexed, and thus stop the horse getting at its legs.

WOUNDS AND SCRATCHES

The most important point when treating wounds is to keep them clean and dry. Whether a wound is best stitched or not is a decision for the vet.

? MY HORSE HAS CUT ITS LEG ON WIRE: MIGHT IT NEED STITCHING?

When a wound heals after being stitched, this is known as 'first intention' healing, and the resulting scar is very small. For this to occur, the wound must be clean and fresh when it is sutured. Unstitched wounds (or stitched wounds that 'break down') heal by what it known as 'second intention' healing. In this process, the deepest and outermost parts of the wound heal first. The repair begins at the edges, and the initial blood clot (later to form the scab) is replaced by new tissue (granulation tissue) growing in from the borders of the wound. New skin later grows in from the edges of the wound to cover the granulation tissue. Whether a wound is best stitched or not is a decision for the vet, and depends on how fresh it is, its depth, and its position on the horse (wounds on the head and body heal well; those on the limbs tend to be less successful). If the wound is fresh, it is best not to touch it at all before the vet arrives (other than trying to prevent it from becoming contaminated). It should not be washed with disinfectants because these can damage the tissues and reduce healing capacity. Clean, lukewarm water can, however, be used to clean the wound, and lint or gamgee applied to a leg wound under a bandage to keep it clean until the vet comes.

If the wound is not fresh (12-24 hours old), it will already be contaminated with bacteria from the skin; there is little point in stitching it for that would only seal infection inside, and it would 'break down' later. Most infected wounds heal very well by 'second intention' healing.

? HOW SHOULD I TREAT DEEP OR INFECTED WOUNDS?

Contaminated wounds can be cleaned and washed with cotton wool and a solution of mild medical disinfectant. A suitable antibiotic ointment (ask a vet) should be applied to the area, under gamgee and a bandage. The ointment not only reduces infection and helps healing, but stops a lint or gamgee dressing sticking to the wound. Likewise, squares of paraffin gauze are useful dressings under lint and bandages for the same reason. Deeper wounds on certain parts of the limbs are likely to involve damage to important inner structures. The deep flexor tendon is sometimes severed by wire wounds at the back of the pastern; the superficial flexor tendons can be cut by over-reach wounds or by being 'struck into' from behind above the fetlock. These are serious injuries – it may even be necessary to stitch together tendons that have been completely severed. You should call a vet immediately if deeper structures appear to be damaged.

Remember, too, that tetanus can develop following a wound. If your horse is not vaccinated against this disease, it *must* have an injection of tetanus antitoxin each time it suffers a wound to give it temporary protection.

? HOW DOES THE HEALING OF WOUNDS IN HORSES DIFFER FROM HEALING IN OTHER ANIMALS?

Wound healing in horses differs from that of other species in that the granulation tissue, which is formed as part of the normal process of 'second intention' healing, tends to form in far larger amounts than is needed. Various factors such as water, irritant powders and ointments also encourage the formation of extra granulation tissue. All this superfluous tissue grows out beyond the normal level of the skin surface, on which it forms a bright pink lump that is commonly called 'proud flesh' by horsemen. Although wounds in other species normally heal better when exposed to air and left unbandaged, those in horses tend to form 'proud flesh' unless pressure is applied to them. It is necessary, therefore, to keep a pressure bandage on limb wounds on horses while they are healing. This may take at least two to three weeks, or even longer for large wounds. Wounds should not be washed during healing: wetting can by itself stimulate proud flesh formation. It is usually adequate to wipe away any discharges with a dry or slightly moistened piece of lint or gamgee when changing the dressings each day. If the bandage is removed too soon, or if a bandage has not been used at all, it will be necessary to remove the resulting lump of proud flesh because new skin will be unable to grow out over the surface to cover it: the proud flesh must be brought back to skin level. This is usually done by chemical cautery – 'burning' it back, using an irritant solution containing copper sulphate. Proud flesh has no nerve supply, and a few days' application of the solution is normally sufficient for the purpose. You should consult a vet for a suitable lotion and directions for its use. Rarely, it may be necessary to remove large pieces of proud flesh either by conventional surgery or by cryosurgery (excision of the tissue by freezing it).

A wound on the outside of the hock showing excess granulation tissue or 'proud flesh' *(left)*. This might require veterinary treatment to remove the excess tissue before the skin can grow satisfactorily over the wound.

MY HORSE HAS A PUNCTURE WOUND: HOW SHOULD I TREAT IT?

Puncture wounds should never be stitched; they should be cleaned, dressed, and allowed to drain so that they can heal from inside first. If foreign material (such as a thorn or needle) is present in the wound, it will go on discharging for many weeks, or even months, unless the object is removed. Indeed, if there is the possibility, at the time of injury, that something is in the wound you should call the vet to investigate and remove it. Likewise, if the injury continues even after treatment to discharge for several weeks, you should also seek help. This, like many other types of wound, can be serious if a joint or tendon sheath is punctured. In a flesh wound, viscous straw-coloured tendon-sheath fluid or joint lubrication fluid (synovial fluid) can be seen; but in longer-standing wounds, which may be oozing serum, these can be harder to distinguish. If there is any suspicion that either a joint or tendon sheath is involved in a wound, call the vet immediately. Extensive treatment with antibiotics is necessary at least, and possibly surgical repair of the punctured joint capsule or tendon sheath.

Puncture wounds should be cleaned thoroughly, daily, with a dilute salt solution (one teaspoon of salt to half a litre – one pint – of warm water) or a dilute antiseptic solution. An old plastic syringe is useful to squirt the solution into the hole, to keep it clean, and more importantly, to keep it open. A solution of hydrogen peroxide (one part to ten parts of warm water) is very useful for cleaning up this form of wound. Poulticing is usually required a day or two after injury, to help draw out pus resulting from infection – which nearly always occurs. Such treatment is especially necessary for puncture wounds in the sole of the foot.

SHOULD I APPLY POWDER OR OINTMENT TO A SCRATCH OR WOUND?

Specially prepared powders containing antibiotics can be useful for application to fresh cuts or wounds to prevent infection. However, some powders can cause irritation which may stimulate the formation of proud flesh (see previous question). Sulphalinamide powder, once ˙popularly used as an antibiotic, is thought to have this property and is now not recommended for use on horses. Aerosol sprays containing antibiotics can help to dry up small wounds when fresh, but horses frequently resent aerosols – particularly around their legs.

When wounds are situated on parts of the body that are continually being moved (the back of the pastern, for instance), applying antibiotic powders or sprays dries the wound, but only makes it more likely to crack again when the animal moves, thus making matters worse. In this case, it is preferable to apply an antibiotic ointment (on a piece of lint or gamgee) and to bandage it in place. Similarly, ointments are preferable for abrasions, to prevent drying and cracking. Pus coming from infected wounds or abcesses can 'burn' and cause damage (in the form of dermatitis) in healthy skin near by. This can be prevented by putting a protective layer of a soothing ointment (such as zinc and castor oil ointment) on the skin around and especially below the source of the discharge.

MY HORSE HAS BEEN KICKED: HOW SHOULD I TREAT THE BRUISING?

Injuries from kicks or blows cause bruising of the muscular tissue below the skin (contusion). For the first 24 hours, administration of cold treatment – ice-packs, cold hosing or cold kaolin poultices – should help stop internal haemorrhage and reduce inflammation. After this, application of warm treatment – hot fomentation and warm poultices – for two to three days may help reduce the bruising and promote healing. If the skin has been broken, treatment as for a wound must be given.

First Aid. Applying antibiotic powder to a wound on the pastern (above).

FIRST AID

All horse owners should have some basic first-aid equipment, and know how to use it.

IS A FIRST-AID BOX NECESSARY?

All horse-owners need to have some basic first-aid equipment, and more importantly, to know where to find it in an emergency. The ideal solution is to have a first-aid box which can be kept in a medicines cupboard in the tack room. This can then be taken to competitions, and can always travel with the horse in a horsebox or trailer.

WHAT SHOULD IT CONTAIN?

A first-aid kit should contain: a pair of surgical scissors; two 8-cm (3-inch) crêpe or elastic bandages; two 8-cm (3-inch) Elastoplast rolls; four 8-cm (3-inch) gauze or calico bandages; one roll of cotton wool; one roll of gamgee tissue; one packet of lint; a tin of paraffin gauze dressing; antibiotic powder in a puffer; antibiotic aerosol spray; a surgical disinfectant; and a thermometer. This kit represents the basic requirements for emergencies when away from the stable with the horse. In addition to this travelling kit, a more comprehensive range of medicines and

equipment should be kept in the tack room. This should include: some form of poultice; a cough linctus or electuary; worm powders or pastes; hydrogen peroxide; table salt; Epsom or Glauber's salts; methylated spirit or witch hazel; lead or other astringent lotion; surgical disinfectant; a general disinfectant (for the building and equipment); four 10-cm (4-inch) woollen stable bandages or 7.5-cm (3-inch) nylon or elastic exercise bandages; and a clean bucket.

MY HORSE OBJECTS TO HAVING ITS WOUND DRESSED: HOW CAN I RESTRAIN IT?

Horses often object to procedures — such as receiving injections or other medical treatment, clipping, or shoeing — and restraint applied in the proper manner can save injury, to either owner or horse. When handling any horse during a procedure that could possibly upset it, a bridle should always be used. This gives much greater control than a head collar and leading rein. If the animal has been well handled, getting an assistant to lift and hold up a foreleg is the simplest way of restraining it, and also helps prevent it from moving unnecessarily. This is useful when taking the temperature, or when examining sensitive areas, such as the groin. It is important that the horse does not lean on

the person holding the leg up because, by using him or her as a support, it can still kick effectively with its hind leg! When a horse will not stand still, or is fooling about, it can often be made to desist by taking a firm grip of a loose fold of skin in the middle of the neck and pinching it in a clenched fist. This is useful for quick procedures, such as giving an injection, but cannot be used for any length of time because of the difficulty in maintaining sufficient grip. For foals and yearlings that need to be restrained, catching hold of an ear is effective. This must be done quickly, and the ear twisted. It can, for instance, be useful when worming unhandled youngsters.

CAN A 'TWITCH' BE HELPFUL?

On no account should a 'twitch' be put on a horse's ear: a 'twitch' should be applied only to the muzzle (the upper lip). Doing this can sometimes be the only safe way of restraining an awkward horse for treatment.

When kindness and deception fail, a twitch applied quickly can be very helpful, and usually makes a horse stand still without causing it distress — in fact it could even produce some relief of pain by pressure on an acupuncture point in the area.

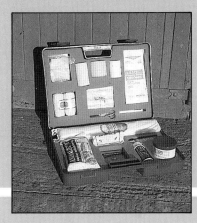

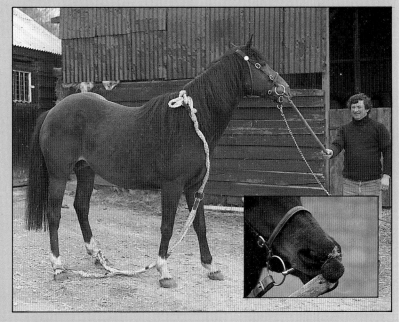

Restraint. A twitch has been applied to prevent the horse rearing and plunging whilst it receives necessary attention

(above). On rare occasions, hindleg hobbles are used to prevent kicking. *(Inset)* Close-up of twitch applied correctly.

EMERGENCY

It is not always possible for the vet to come at once when an accident occurs. When a horse is bleeding, suffering from colic, or choking, the owner must know how to administer treatment fast, whilst keeping a cool head

MY HORSE HAS GOT ITSELF 'CAST': WHAT SHOULD I DO?

A horse is said to be 'cast' when it has somehow got itself into a position in the stable with its feet lying against the wall, in such a way that it is unable to get up. Commonly, this is the result of a horse's rolling in a small stable (sometimes as a result of colic). When it is discovered in this predicament, a horse should be approached quietly to avoid further struggling and possible injury. An assistant is required to restrain the head (probably by kneeling on its neck) while you attach two long ropes (if possible cart-ropes with hobbles; if not, lungeing reins will do) around the pasterns of the fore and hind legs that are against the wall. By slow, steady pressure, it should be possible to pull the horse over. It can then get up, and should then be closely examined for signs of injury to its back or limbs.

MY HORSE IS BLEEDING: WHAT SHALL I DO?

Control of bleeding is the first stage of treatment of fresh wounds. Applying pressure, by means of lint gauze held in the hand, is usually sufficient when blood is oozing from minor wounds. Cotton wool should not be used for this purpose because small particles tend to be retained in the wound. For more extensive bleeding, a pressure bandage is desirable. When arteries are cut, the blood is bright red and comes in spurts. In this case, a lint or gamgee gauze dressing should be placed over the area, followed by the application of a tight elastic or crêpe bandage. The pressure not only slows down the bleeding but also keeps the edges of the wound together. Movement stimulates blood flow. The horse should be kept calm and tied up in the stable to avoid movement until the vet arrives. When bleeding occurs in an area that cannot be bandaged — for example, a nose-bleed — applying ice-packs or cold hosing may help. Tourniquets are not advisable for horses: they often make matters worse.

IF THE STABLE CAUGHT FIRE, TRAPPING MY HORSE, WHAT SHOULD I DO?

Horses are terrified of fire and can refuse to be led through an open stable door to safety. In this case, some form of blindfold may be needed. A coat or riding jacket can be used, but a dampened stable rubber or similar large cloth (fixed under the browband of a bridle and hanging down the horse's face) may prevent eye irritation and inhalation of smoke while the animal is being led to safety.

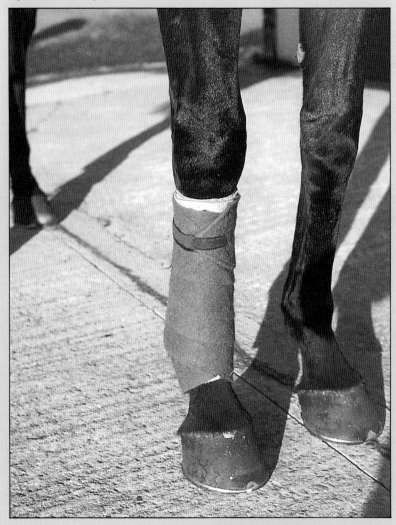

Applying a bandage after treating an infected wound *(right)*. The bandage is applied from the coronet up to the knee to exert pressure and prevent the leg swelling.

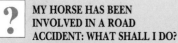

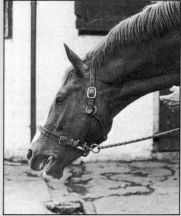

A horse with acute **colic** will roll repeatedly if left on its own; this can cause a fatal twisted gut *(left* and *below)*. The horse should be kept standing until the vet arrives.

A horse choking. It is quite rare for a horse to choke, but it can be very dangerous *(above)*. Veterinary help should be called immediately.

? MY HORSE HAS BEEN INVOLVED IN A ROAD ACCIDENT: WHAT SHALL I DO?

It is important not to move a horse until the full extent of an injury has been assessed — a vet is the best person to do this. Sometimes a hairline fracture of a bone occurs at the site of the original injury, and a full fracture occurs only when the horse is moved. If the horse is down on the ground and seems to have suffered major injury (perhaps a broken leg), it is important to prevent it from struggling to get to its feet. To do this, its head must be restrained. This can be done by lying across or sitting on its neck just below the head. A coat or similar blindfold may help calm the animal until help arrives. When moving or transporting a horse that may have a broken leg or other serious limb problem (ruptured tendon, or whatever) it is very important to provide as much support as possible to the injured leg. This can best be done by using whole rolls of cotton wool wrapped all the way round the limb and incorporating one or more 10-cm (three-inch) crêpe or elastic bandages which can be pulled tight to compress the cotton wool. By this means a thick 15-20cm (six-to eight-inch) solid supporting cotton wool cast can be produced. Incorporating splints of wood into this helps give the added strength required for temporary support for a fracture.

? MY HORSE IS CHOKING: WHAT SHOULD I DO?

Choking can occur for a variety of reasons. A vet should be called immediately because there is the danger of 'inhalation' pneumonia. A small drink — 0.65 litres (1 pint) of water — can be given to try to clear it, but more should not be allowed (remove the water bucket): larger amounts may be inhaled.

? MY HORSE IS SHOWING SIGNS OF COLIC: WHAT SHOULD I DO?

Except in very mild cases, colic should always be treated as an emergency. By getting down and rolling, horses are capable of twisting one part of their intestines around another — which can have fatal consequences. So the sooner a vet is called to give painkilling injections, the better. If a horse is discovered rolling in pain, it should be got to its feet, bridled, and walked quietly round (on grass) until the vet arrives. In the meantime, get an assistant to make up a deep straw bed in the largest available stable. Should the animal repeatedly try to throw itself to the ground, it is safer to return it to the prepared box rather than risk injury. If it has sweated up and is now cooling down, rugs may be required to keep it warm.

? MY HORSE APPEARS TO BE SHOCKED: WHAT CAN I DO BEFORE THE VET ARRIVES?

Horses suffering from shock are depressed and feel cold to the touch (and may have a subnormal temperature). They often have a weak, but rapid, pulse and a bluish tinge to the visible mucous membranes (in the gums, nostrils and eyes). Shock can result from injury, such as internal haemorrhage, or from acute illness, such as diarrhoea, colic, laminitis or azoturia. While awaiting professional help it is important to get the animal into a warm stable and put on rugs and stable bandages.

GENERAL MEDICAL CARE

8

Anyone looking after a horse or pony must be able to recognize the signs that indicate an animal is ill or injured. Horses suffer from a bewildering variety of ailments, many of which have no human equivalent and have names that are often of little help in understanding the diseases and what causes them. It is not necessary for an owner to have a detailed knowledge of all these conditions, and how to diagnose and treat them. That is a vet's job. However, a knowledge of what can go wrong, and the signs that may be shown, can make the detection of illness and injury in a horse easier. This should prevent unnecessary delay in seeking professional help and obtaining treatment, and thus help to speed recovery.

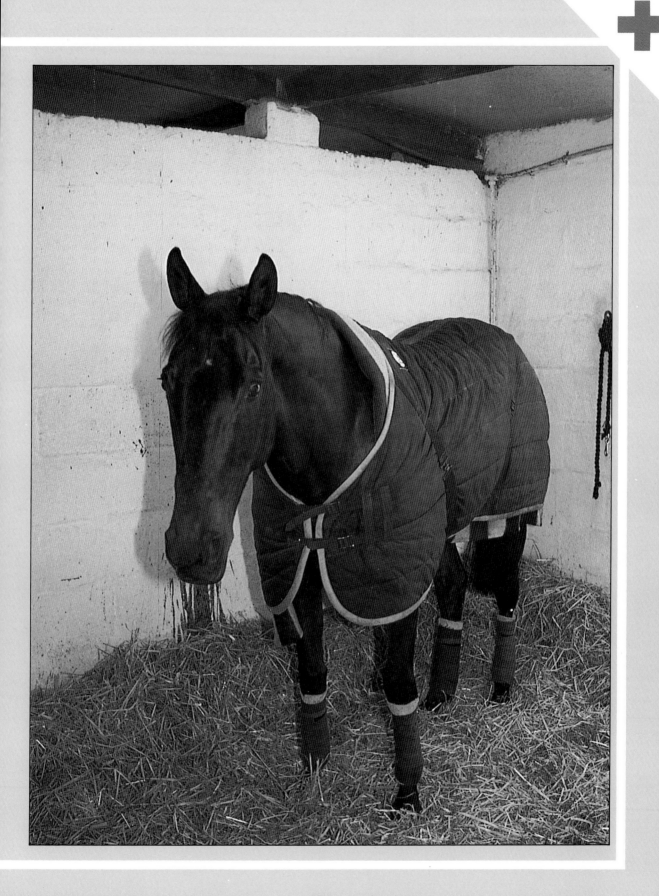

SKIN PROBLEMS

Horses are prone to a number of skin problems and complaints. Many of these can be avoided by good management. When they do occur, early diagnosis and treatment are

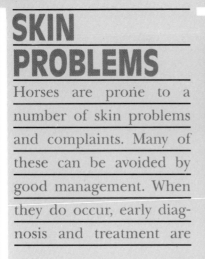

? **MY HORSE HAS A LUMP IN ITS SKIN UNDERNEATH THE SADDLE AREA: WHAT CAN THIS BE, AND WHAT SHOULD I DO ABOUT IT?**

In the past, skin swellings under the saddle area were known as saddle galls or, less often, 'sitfasts'. There are several possible causes of such swellings, and all may cause problems with pressure from the saddle. One of the commonest swellings occurs in the middle of the back. This is usually hard and is a result of pressure from an ill-fitting saddle over the vertebrae. Hard lumps in the skin, known collectively as nodular skin disease, occur on many parts of a horse's body, and are a result of an allergic reaction to the bites of flies (particularly the horse-fly and the stable-fly). Such hard swellings under the saddle may also cause problems. Bacterial skin infection — acne — can also occur under the saddle, and may be spread from horse to horse via tack and grooming kit. Saddle pressure on these lesions can result in nasty sores.

Another less common problem on horses' backs is caused by the emergence of warble fly larvae through the skin. Warbles more commonly affect cattle, but when horses are grazed with cattle, a few of these flies may attack them. The eggs are swallowed by the horse, and during the next year, migrate from the intestines to emerge through the back. Occasionally the grubs do not emerge but die, and remain just beneath the skin surface where they can provoke a considerable reaction. With all skin problems under the saddle it is important to protect the affected area from saddle pressure until it has healed. This is done either by resting the horse (when there are

many small spots — such as with acne), or by using a foam pad in which an area has been cut away to relieve pressure over the sore spot. Specific treatments may also be required — antibiotic ointments and antibiotic injections for bacterial infections; poulticing or hot fomentation for warbles or back abscesses; applications of witch hazel for sore backs and girth and saddle galls. Surgery may be needed to remove particularly troublesome lesions of nodular skin disease or hard swellings over the backbone.

? **MY HORSE'S HAIR IS FALLING OUT: HAS IT GOT LICE?**

Lice are quite common in horses in winter. Two kinds of lice affect horses, a biting louse and a sucking louse. Both of these provoke intense itching and the animal will rub its skin (particularly over the head and neck), producing large bald areas in the process. Lice can usually be seen moving in the coat: they are white or yellowish in colour and are about the size of a pin-head. The only other condition which could be confused with lice is 'sweet itch', but because this only occurs in the summer, there is little likelihood of confusion. There are special insecticidal shampoos and louse powders available. These are very effective in eliminating the lice but it takes some time thereafter for the hair to grow.

White hairs on the withers and shoulders indicate previous **pressure sores** ('saddle galls') caused by a badly fitting saddle. The sores *(above)* have healed following rest and treatment, and surgical spirit is now being applied daily to harden the skin. A foam pad or 'numnah' under a correctly fitting saddle provides extra protection. *(Above right)* Close-up of shoulder of the same animal showing the area of skin damage.

The dull scaly coat on the horse's back *(below)* is the result of **lice infestation**. Lice are frequently found on horses in winter and early spring. They can be prevented by routine dusting with an anti-parasite louse powder.

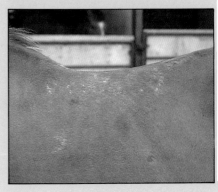

lesions usually first appear under areas where the skin has been rubbed by tack. These are small, rounded patches, which gradually enlarge and show signs of hair loss. Within this area small fluid-filled vesicles may form, which may rupture and form scabs, sometimes causing irritation at this stage. Various topical fungicides are effective for treating lesions, and it is important also to disinfect and clean tack and grooming equipment with a disinfectant that kills fungi (many disinfectants do not do this – ask your vet to recommend a suitable one). If the horse is very badly infected with many areas of hair loss, it may be helpful to treat the animal with a systemic fungicide (such as griseofulvin) by mouth. (This drug, however, should not be given to pregnant mares.)

WHAT CAUSES 'HEEL-BUG'?

The larval stages of the harvest mite (*Trombiculum autumnalis*) are red in colour and suck blood from many species of animal (including humans). They appear in late summer and autumn, particularly on chalky soils and in cornfields. At harvest these mites become incorporated into straw bales, and bedding contaminated with these minute larvae is the cause of 'heel-bug' in horses. If the red mites are detected, or suspected from intense irritation around the legs, it is essential to change the bedding, and the legs can also be washed in an insecticidal shampoo. Application of benzyl benzoate may help to kill any larvae on the limbs and soothes the skin.

DO HORSES SUFFER FROM SKIN GROWTHS, AND ARE THERE ANY TO BE WORRIED ABOUT?

Skin growths in horses are quite common, but, fortunately, the majority of these are benign. Very few are malignant, but it is probably as well to get your vet to check any growths, particularly if they grow rapidly or are found in areas where they are likely to cause problems – around the eye, in the groin, or under tack. Many grey horses have growths – usually around the tail and anus, especially when they are older. Although these growths (melanomas) are usually benign, they can, rarely, spread to internal organs with serious consequences. As with other skin growths, it is wise to let your vet examine them; take his advice as to whether they should be left or removed.

WHAT IS 'SWEET ITCH'?

'Sweet itch' is a condition affecting horses' manes and tails. Affected animals suffer intense irritation, and most of the damage is done by the horse rubbing itself to relieve the condition. This can lead to loss of hair and raw sores on the root of the tail and on the withers and up the neck.

The name of this disease is derived from 'sweat itch' – although the reason for this is obscure because the disease does not occur on the parts of the body where the horse sweats. Only certain individual horses (about 2% of the equine population) suffer from this condition, which is caused by an allergy to the saliva of certain biting midges. The midges involved are members of various species of *Culicoides*, which are only active in the summer months. These are more common in wet and marshy areas and are only active around dusk when they emerge to feed. Once signs of sweet itch develop, it is a difficult condition to treat, often requiring injections of cortisone or antihistamines to break the itch-scratch cycle. Various lotions may be helpful in soothing the skin and preventing it from drying and cracking – benzyl benzoate is frequently used for this purpose. However, once it is realized that a horse suffers from this allergy, the best course of action is to try to prevent the condition from occurring in future years by taking vigorous measures to prevent the horse from being bitten by midges. Each summer the horse should be removed from grazing daily at 4 p.m. and stabled until the following morning. Keeping the horse stabled during the time when the midges are active is the only way of preventing sweet itch. The stable windows and any opening above the stable door should be fitted with fine-mesh wire grilles, and fly papers and slow-release insecticidal chemical strips should be fitted inside the stable. The horse might also be treated with fly repellants.

It seems likely that in the future it may be possible to produce a vaccine to desensitise animals that are allergic. There is some evidence too that this allergy can be inherited, and the wisdom of breeding from affected animals is thus questionable.

CAN HORSES GET RINGWORM FROM CATTLE?

The most common form of ringworm in horses is caused by the fungus *Trichophytosis equinum*. Infection with this organism is acquired only from other horses. However, if horses are housed in buildings that have been occupied by calves or cattle infected with ringworm they can contract infection with the bovine form of the disease (*Trichophytosis verrucosum*). Ringworm in horses is usually acquired by direct contact with another infected horse or through infected tack or grooming kit. There is an incubation period of five to ten days, and

WHAT IS THE MORE COMMON FORM OF SKIN GROWTHS?

Sarcoids are perhaps the commonest skin growths in horses, and there is still some argument as to whether these are true tumours. Recent research has shown that animals who suffer from wart virus infections when young are much more likely to suffer from sarcoids when they are older. It is thought that some breakdown in the immune system may be responsible for the subsequent development of these growths. Horses with sarcoids develop one or more — often several — tumour-like masses on the body, frequently on the chest, lower abdomen and in the groin (particularly on the sheath of males). They are frequently knocked and may bleed as a result, and can become infected and ulcerated. Occasionally, a growth develops on a stalk and it may be possible to tie a ligature around this which then eventually drops off. Rarely, sarcoids can reach enormous size — weighing up to 9 kilograms (20 lbs). Chemicals have little effect on these growths but they can be removed by surgery under local or general anaesthesia. However, there is a great tendency for growths to recur, or for fresh growths to appear elsewhere on the body after conventional surgery. For this reason sarcoids are frequently removed by the freezing technique, known as cryosurgery. This is more effective; there are fewer problems with recurrence when this method is used. In cryosurgery the horse is anaesthetized while metal probes attached to some means of freezing to a very low temperature (eg liquid nitrogen) is applied to the growth; at the same time a temperature-measuring device (thermocouple) is placed in surrounding healthy tissue to make sure that this is not frozen. The growth is frozen, then allowed to thaw before being frozen a second time. There is no pain following this form of surgery: the tissue gradually withers and the wound heals in the same way as any other wound. Because these growths are likely to spread and recur it is wise to avoid buying a horse which has them, wherever possible.

A skin growth in the corner of the eye *(above)*. This is in a particularly troublesome position, and will cause irritation, weeping, and problems with flies. Surgery is often necessary to remove growths of this kind.

The condition *(right)* is known as **cracked heels** – a moist eczema caused by repeated wetting of the heels, resulting in a thickening and cracking of the fold of skin at the back of the pastern joint.

WHAT ARE RAIN SCALD AND MUD FEVER?

Rain scald and mud fever are caused by a bacterium, *Dermatophilus congolensis.* Unlike many other bacteria, this organism forms spores capable of surviving for many years on horses' coats and in their environment. The organism is also one of a type known as anaerobes, which only grow in the absence of oxygen. For the spores to germinate, continual wetting is required. Thus in wet weather 'rain scald' develops on the thick, matted coats on the hindquarters of horses at grass in winter. The infection develops in the skin under the matted hair, beneath a thick scab. The matted hair and scab can be removed, and resemble the hairs on a paintbrush – they are sometimes called paintbrush lesions for this reason. A similar infection can occur on the legs when they are wet and caked with mud, most usually on the lower cannon area, where it is known as 'mud fever'. Because the bacterial organism causing the infection cannot survive in the presence of oxygen, removing the scabs (where this is sensible – that is, without producing huge raw areas) exposes the pus in the underlying sores to the air. This in itself is usually sufficient to dry them up, but dabbing on a solution of 1% potash alum speeds up the process. The scabs are usually best plucked off the quarters, but it is easier to wash and scrub them off the legs, using an antibacterial shampoo (such as a surgical scrub). In severe infections, or when mud fever causes the legs to swell, it may be necessary to give a course of antibiotics. Above all, it is necessary to stable the horse, and to keep it clean and dry until the infection has cleared up.

WHAT CAUSES 'CRACKED HEELS'?

Cracked heels are the result of dermatitis on the back of the pastern. They occur in wet, cold weather, and may be due to irritation of the skin by dirt. More than one leg is usually involved, and legs with white markings seem to be more prone to this condition. Heels usually become cracked as a result of chilling horses' legs by washing them to clean them in cold weather. This should be avoided: allow mud and dirt to dry, and then brush them off, rather than wetting the leg. In mud fever, the skin above the heels becomes reddened, tender, and scaly. Later, small fluid-filled vesicles develop, which rupture, and cracks appear across the skin. These may ooze serum which dries on the surface, but the wounds often crack again when the animal moves. The heels can become very sore, and the horse may become lame. Secondary bacterial infection is common, and may spread up the leg. The soreness can be so great that a hind limb is snatched up, and the condition could possibly be confused with stringhalt. Treatment involves washing affected legs thoroughly with a medicated shampoo to remove any pus or serum exudate, and drying them as much as possible, before applying a soothing ointment (such as zinc and castor oil ointment). This should be applied on a piece of gamgee or lint and bandaged to the leg. It may be necessary to apply such treatment for some time, but this should prevent repeated cracking of the wound, which would otherwise delay its healing. Application of ointment to the heels in wet weather, and use of stable bandages at night, might help to prevent this problem from occurring at all.

DO HORSES SUFFER FROM SKIN ALLERGIES?

There has been little scientific investigation of skin allergies in horses, but allergic reactions are not uncommon. These take the form of raised wheals (fluid-filled plaques) in the skin, which are small at first but may enlarge rapidly. It is thought that most cases of such reaction occur after the horse has eaten something to which it is allergic – a particular plant when at grass, or in hay. Sometimes alternatively, however, it appears that something in bedding, particularly straw, is causing the horse to react, and a change of bedding material is necessary. Horses can also be allergic to particular drugs, and may show skin reactions, occasionally also accompanied by lung and circulatory effects which can be serious. The skin wheals may be intensely itchy, and the condition – which is known as urticaria – usually subsides fairly quickly without treatment. However, if a large area of the body is involved, or the horse shows any sign of distress (such as heavy breathing, restlessness or unsteadiness on its feet) the vet should be called immediately. Anti-histamine injections can help resolve the skin problems, but if a hypersensitivity reaction to a drug is involved, treatment for the shock that can result may be needed urgently. At the same time it should be remembered that horses can be allergic to penicillin and other antibiotics, in the same way as human beings. Giving them to hypersensitive horses can have serious consequences and, if there is any suspicion that your horse has previously shown an adverse reaction to any particular drug, you should inform the vet of this·

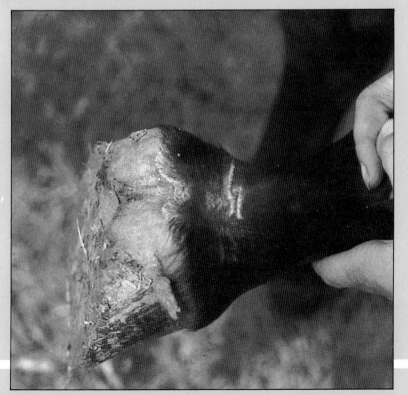

EYE PROBLEMS

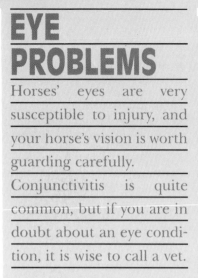

Horses' eyes are very susceptible to injury, and your horse's vision is worth guarding carefully. Conjunctivitis is quite common, but if you are in doubt about an eye condition, it is wise to call a vet.

IS THERE ANY FUNDAMENTAL DIFFERENCE BETWEEN EQUINE AND HUMAN VISUAL APPARATUS?

Like cats and dogs, but unlike humans, horses have a third eyelid. This comes across the eye from the front corner (the inner canthus) to protect it. Rarely, cysts may develop within this eyelid, and it is a not uncommon site for malignant tumours. Any swelling in this eyelid should be treated with suspicion, and veterinary advice sought. The third eyelid can be stitched across the eye, and can, in this way, provide marvellous support and protection for wounds to the front of the eyeball (the cornea) while they heal.

MY HORSE HAS CONJUNCTIVITIS: HOW CAN I TREAT IT?

Conjunctivitis is quite common and often results when dust, or particles of hay, bedding or other foreign material enters the eye. The eyelid mucous membranes become reddened and swollen, and the eye waters excessively. The condition should respond rapidly to treatment with antibiotic eye ointments and bathing of the eyelid with cotton wool and warm water to remove any accumulated matter. If the eye is particularly itchy, it may be necessary to tie the horse up 'short' for a while, until the ointment has a chance to relieve the irritation, in order to prevent the horse from rubbing the eye on its foreleg and making it worse.

Conjunctivitis can also occur with injuries to the cornea, and with periodic ophthalmia.

HOW SHOULD EYE INJURIES BE TREATED?

Horses' eyes, being particularly prominent on the side of the head, are very susceptible to injury. Overhanging branches and stalks frequently damage the front surface of the eyeball (the cornea), and injuries vary from a superficial scratch to complete penetration and rupture of the eyeball. Such injuries are always serious, and the vet should be consulted straight away. If the cornea is completely ruptured, it is feasible to suture and repair it — but it must be done immediately — unless the eye is infected. A superficial corneal wound or injury (frequently found when clods of earth are thrown up by horses in front) always requires treatment with antibiotic eye ointments. There are always bacteria present on the eye, and these rapidly invade damaged corneal tissue, and may cause ulcers. If an ulcer becomes deep, or does not respond to antibiotic treatment, it may be necessary to cauterize it, under general anaesthesia, to help it heal. Eye injuries are often very painful, and the pain is made worse by spasm of the iris (causing a contracted pupil). Giving eyedrops containing a mydriatic (such as atropine) to dilate the pupil should relieve much of the pain. Healing of corneal wounds is slow because there are no blood vessels in the cornea. These must 'grow' in from the side of the eye, and can be seen doing so: this is a sign of healing and nothing to worry about. After healing, a permanent scar, visible as a white spot, will remain. Unless very large, these do not usually affect a horse's vision.

WHAT ARE THE BLACK LUMPS ON A HORSE'S PUPIL — ARE THEY NORMAL?

Horses differ from other species of animal in that they have several (often three) black masses on the upper edge of their irises (the circular brown area surrounding the pupil). These, called corpora nigra, are easily visible from a distance, and are completely normal.

DO HORSES GET CATARACTS?

A cataract is an opacity of the crystalline lens of the eye which, to a varying degree, obscures vision. They are quite common in older horses, but fortunately they normally affect only part of the lens and do not completely obsure vision. The pupil of affected animals appears white and shining ('glassy'), but examination with an ophthalmoscope is necessary to detect the full extent of changes in the lens. It is perfectly feasible to treat a cataract by removing the lens, as is done in humans, but this is seldom done with horses because a horse is unable to focus and has problems in bright light afterwards. Old horses in familiar surroundings often manage to cope extraordinarily well with failing eyesight, and it is not unknown for an animal which is almost completely blind to be able to jump in response to commands from its usual rider when it is practically incapable of finding its way around in strange surroundings on its own. While this practice is obviously dangerous, it does indicate how clever horses are in using their other senses, and the difficulty in interpreting what a horse can or cannot see in familiar surroundings. There are several totally blind Thoroughbred mares in the Stud Book. These manage very well in their own home environment, and are helped by fitting a bell to their foals, which prevents them from fretting.

WHAT IS PERIODIC OPHTHALMIA?

This is the name given to a recurring inflammation (recurrent uveitis) of the inner structures of the eye, and the iris in particular. It is one of the commonest causes of blindness in horses. Its own cause is unknown, but the condition may well be the result of an auto-immune reaction, stimulated by previous exposure to some infectious process. In the United States it has sometimes been associated with infection with a bacterium called *Leptospira pomona*, but this does not appear to be the case in Britain, where this organism has not been found. Sometimes uveitis may develop following a respiratory disease, but more often there is no history of any recent infection. The first signs are usually conjunctivitis and watering of one eye. The pupil is often constricted and may be stuck to the cornea and unable to dilate. This can be checked by taking the horse into a darkened box, and comparing the pupil with the unaffected eye. The inflammatory reaction produces a large amount of pus within the eye, which commonly accumulates at the bottom of the front of the eye — a thick, yellow appearance in front of the lower edge of the iris (below the pupil) should thus be viewed with suspicion and reported immediately to the vet. Treatment involves using eyedrops to

Applying **eye ointment** to the corner of the eye *(above)*. They eyelids have been drawn back as far as possible, to allow the ointment to be placed well into the eye. Great care should be taken whilst doing this; horses' eyes are as sensitive as humans'.

dilate the pupil and relieve the pain, and cortisone eyedrops and injections to reduce the inflammatory reaction. This condition is serious. It tends to recur in the affected eye, and often later goes on to involve the other eye. The inflammation damages the lens (causing cataracts), and may cause adhesions between the iris and the cornea, or the lens. This may prevent the pupil dilating and contracting. The end result is often a loss of sight in one or both eyes. The sooner this condition is detected and treated the better. It is sometimes difficult to distinguish the early signs from conjunctivitis, and it is important that all horse-owners should be aware of the possibility of periodic ophthalmia, and its seriousness. They should err on the side of caution, and have their horses' eyes examined by a vet in case the condition is developing.

? MY HORSE'S EYE IS WEEPING: SHOULD I BATHE IT?

Horses' eyes frequently water in summer, at grass, when they are troubled by flies. It is helpful to wash away any accumulated discharges around the eye which may be attracting flies, using cotton wool or a sponge and plain water only. In the stable, washing the eyes with a damp sponge removes similar material, and prevents hay and bedding from sticking to it. If one eye is weeping excessively, this is usually a sign that something is wrong, particularly if the eyelid is closed. The eye may have been injured, or a foreign object such as a hay-seed or fly may be irritating it. Rarely, the tear duct may be blocked. Excess watering can be a sign of periodic oph'thalmia. If there is a thick or pussy discharge, it is most likely that the animal has conjunctivitis, and you should seek veterinary advice. Detailed examination of the eyes is difficult because horses are very sensitive about them. If the eye is closed, or there is any doubt whether it could be injured, it is better to have it examined by a vet straight away.

FEEDING PROBLEMS

Like people, horses vary considerably as to the extent of their appetite. However, if they leave their feed untouched regularly, it is generally a sign of painful teeth or some digestive disorder.

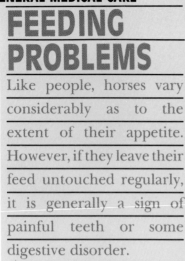

❓ MY HORSE WON'T EAT ITS FOOD: SHOULD I WORRY?

Loss of appetite is frequently a sign of ill-health, and is common in horses with a fever. It can also be associated with some digestive disorders: horses with colic and/or impaction (constipation) may also fail to eat. Sometimes, they are interested in their food but have difficulty eating it, and may even drool saliva, or drop food from the mouth ('quidding'). This is a sign that something is wrong in the mouth, and it is worth while looking inside to see where the problem lies. Not uncommonly, the piece of tissue underneath the tongue which attaches it to the lower jaw (the frenulum) is torn when the tongue goes over the bit. This can be very sore, and makes it very difficult for the horse to eat.

Likewise, cuts inside the mouth from the bit, from sharp teeth or from sharp objects, may cause the same problem. Rarely, sharp objects, such as thorns, may lodge in the gums or cheeks and may also cause puncture wounds. Luckily, the tissues inside the mouth are well supplied with blood vessels and heal quickly, provided they do not become contaminated with food. To help healing, it may be necessary to change to a gruel diet, and to avoid fodder, which tends to trap in puncture wounds. Whereas it should be possible for an owner to spot injuries in the front of the mouth and tongue, problems further back need inspection by a veterinary surgeon using a gag (a metal instrument to hold a horse's mouth open).

A very bad **parrot mouth**. The upper teeth are so far in front of the lower ones that it is impossible for this animal to graze properly.

❓ WHAT DENTAL PROBLEMS MIGHT AFFECT MY HORSE?

If the incisor teeth do not meet properly, the horse's ability to graze may be affected. This occurs in congenital conditions, when the upper incisor teeth are in front of the lower ('parrot mouth'), or when the lower teeth are in front ('undershot'). It is best to avoid buying horses with either of these defects. The first cheek tooth (the 'wolf' tooth) may or may not be present, and can be responsible for cut mouths and sore gums. Wolf teeth are often associated with equitation problems, and are frequently removed for this reason.

Efficient crushing of fodder requires flat, level tables (grinding surfaces) of the molar teeth. Unlike human teeth, those of horses are continually erupting through the gums all the animal's life. Thus, as the surface is worn away, it is being replaced by new tooth from underneath, and the wear from the equivalent tooth in the opposite jaw, should keep the tables level. In practice, wear is often uneven; the sharp outer edges of the upper teeth, and the sharp inner edges of the lower teeth, tend to cut the cheeks and tongue respectively. Also, the tables may be worn irregularly, forming ridges and crevices in which food can accumulate. The latter may allow tooth decay (caries) to develop. Regular rasping helps to keep the cheek teeth tables smooth and level, helps to avoid cuts, and helps ensure efficient crushing of food.

❓ WHEN IS DENTAL SURGERY NECESSARY?

Rarely, infection of the tooth root may occur, forming a tooth root abscess which can cause swelling in the lower jawbone or in the face (below the eye). Such abscesses may respond to antibiotics, but it is

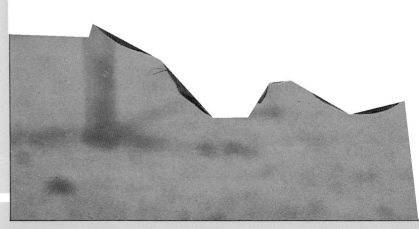

sometimes necessary to remove the tooth to solve the problem. Horses' cheek teeth are very well imbedded in the bone, and although it may be possible to remove lower teeth via the mouth, using dental extractors, removal of upper teeth by this method is usually impossible unless they are loose. It is generally necessary to make a surgical opening into the maxillary sinus (below the eye), and to force the tooth down into the mouth (by pressure on the root from above) to remove an upper cheek tooth.

Dental problems are more common in old horses. If teeth are lost, the opposing crowns cannot be worn but tend to become very long and sharp, and may cut the cheeks or tongue. Likewise, the tables of other teeth may become very irregular in old age. It is sometimes necessary to anaesthetise these animals in order to cut off and rasp sharp crowns. It may also be necessary to remove loose teeth.

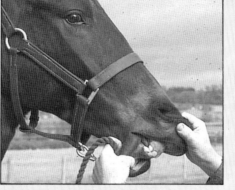

Checking a horse's mouth and teeth.
Routine checks such as this *(above* and *above right)* are essential. Loss of condition, digestive upsets, bad temper and disobedience are frequently a result of dental troubles.

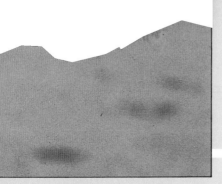

? CAN HORSES VOMIT?
Because of the anatomy of the back of their mouths, horses are incapable of regurgitating food from the stomach into the mouth, and are therefore unable to vomit. In some conditions, when the stomach becomes over-distended, food may pass up the gullet and appear at the nostrils. Such regurgitation is found in very young foals with an obstruction at the exit from the stomach into the intestines (pyloric stenosis), when milk appears at the nostrils. (This must be distinguished from a congenital cleft palate, in which milk from the mouth may enter the nostrils.)

In adult horses, over-distension of the stomach may result in gastric contents coming from the nostrils, commonly following a 'twisted gut' or paralysis of the bowel ('grass sickness'). Food coming from the nostrils is always a sign that something serious is wrong.

? MY HORSE IS SUFFERING FROM CHOKE: WHAT IS LIKELY TO HAVE CAUSED THIS, AND WHAT SHOULD I DO ABOUT IT?
In choke, food becomes lodged in the gullet and is unable to pass on down into the stomach. Affected horses are usually distressed, make repeated swallowing attempts, and may arch the neck in pain. Saliva frequently drools from the mouth, and

food may be regurgitated down the nostrils. The horse also coughs, and there is a considerable risk of food being inhaled into the lungs. If that happens, it may subsequently cause inhalation pneumonia, which develops 12-24 hours later. The most likely objects to lodge in a horse's gullet are large pieces of vegetables (carrots, potatoes), apples (especially wind-falls at grass), concentrate cubes and, especially, sugarbeet pulp that has been insufficiently soaked.

Choke is always an emergency, and a vet should be called immediately. Although a small drink of water — half a litre to a litre (1-2 pints) — may enable the animal to swallow the obstruction there is the danger that the water may instead be inhaled, so the water bucket should be removed from the stable. The vet can usually push the obstruction down into the stomach, using a stomach tube or a fibre-optic endoscope. Small amounts of liquid paraffin, via the stomach tube, help this process. If it is not possible to relieve an obstruction within 24 hours, damage of the gullet lining occurs, with serious consequences. It is then necessary to relieve the choke by surgery.

Choke is often associated with greedy feeders who bolt food down without chewing it. Putting several large stones or bricks in the manger often makes a horse take longer to find and eat its concentrates. By slowing down an animal's feeding rate, and giving smaller feeds more often, the risk of choke is reduced.

Colic – acute pain in the digestive organs – will often make a horse roll in an effort to relieve the pain. It should be prevented from doing this, however, as rolling could twist its gut, which will cause death.

WHAT IS COLIC, AND WHAT ARE THE SIGNS?

Colic is a term used to describe abdominal pain. Although pain can be associated with kidney problems (renal colic), with horses the term is generally used to describe pain that originates in the intestines. A horse's intestines are very well supplied with sensory nerves that respond to the slightest discomfort; an animal may thus make as much fuss about the slightest twinge of indigestion as it does when suffering from a major abdominal catastrophe. It is therefore very difficult for horse-owners to tell whether or not something is seriously amiss when their horses have colic.

The signs of colic vary slightly with the cause, but the animal usually shows loss of appetite, lethargy (sometimes yawning) and general lack of interest in its surroundings. It may look round at its flanks, paw the ground with a fore foot, or kick at its belly with a hind leg. If the pain increases, the horse may frequently get down and roll, in an attempt to obtain relief. It may remain lying on its back with its feet in the air, an attitude that appears to offer more relief (particularly in foals). There may be very hard droppings, no droppings, or diarrhoea, depending on the cause, and the horse may or may not sweat up. In serious cases, signs of shock may become evident — cold extremities, bluish visible mucous membranes and a rapid, thready pulse.

IS COLIC SERIOUS?

Although some causes of abdominal pain, such as impaction (constipation), are unlikely to cause more than temporary discomfort, the signs which the horse may show are indistinguishable from the early signs of serious complaints. In addition, a horse's natural reaction to colic is to roll to relieve the pain. This can itself cause a serious complication — a 'twisted gut' (see below), which if not resolved is fatal.

In many cases of colic, all that is required is pain relief, to prevent the animal from injuring itself and developing complications. The sooner a vet is called to give this the better; in the meantime colic should always be treated as an emergency.

WHAT IS A 'TWISTED GUT'?

Horses' intestines are wrapped in a membrane (mesentery) through which the blood vessels and nerves supplying the bowel run. The membrane is attached to the roof of the abdomen, and within it some parts of the intestines are also firmly attached and fixed in position, although others hang free and can move. The small intestine has the most freedom of movement, and can become twisted around itself and knotted (forming a 'twist'). Bowel trapped in this way is said to be 'strangulated' because its blood supply is cut off. Tissue deprived of blood becomes painful, and this is the cause of the acute pain associated with a twist. At the same time, the internal composition and content of the trapped portion changes: it fills with fluid, and the digestive bacteria normally present produce toxins which are responsible for the signs of shock which develop.

Twists can occur spontaneously as the primary cause of colic (particularly in foals or yearlings), or may be a complication of other forms of colic, brought about by a horse's rolling. Bowel obstruction may also result if a length of intestine works its way through any small hole in the outer membrane and gets trapped through it in an abnormal position. This can happen through natural openings in the mesentery ligaments around the stomach, or through abnormal holes that

have for one reason or another developed in the mesentery. Occasionally, fibrous bands may be present within the abdomen (usually as a result of migrating worm larvae damage). These can become wrapped round the intestines, obstructing it.

Intestinal strangulation can be corrected only by surgery. Regrettably, this is not always successful, but the horse will die of shock unless the obstruction is relieved. The chances of recovery depend entirely on how quickly the strangulation is detected and corrected. If a twist has been present for more than six or eight hours, the outlook becomes much less favourable. It is usually impossible to determine which portion of the bowel is involved until the abdomen is opened. The diagnosis of a twist is not always straightforward, and the use of the more potent painkilling drugs now available tends to make some of the signs far more difficult to observe unless the case is closely monitored. Regular examination of the pulse, listening to abdominal sounds, examination of the intestines by means of a hand inserted into the rectum, blood tests, and possibly tests on samples of peritoneal fluid — all these may be necessary to help establish a positive diagnosis. Failure to respond to painkillers, and signs of shock are suggestive that something is seriously wrong. Surgery, under general anaesthesia, through an opening in the middle of the abdomen, to locate and the large intestine and the caecum — a common site of bowel obstruction — have been observed, during surgery, to have large numbers of tapeworms. There is now increasing circumstantial evidence that tapeworms may be a possible cause of problems at this site.

? WHAT IS SPASMODIC COLIC?

This is a form of recurrent colic, and is probably the most frequent type encountered. The pain is usually sudden in onset, and is accompanied by the normal signs of colic together with a raised pulse and more rapid breathing during painful episodes. The pain normally subsides after a few hours, and returns 6 to 12 hours later. No abnormality is usually detected when listening to the bowel sounds or by rectal examination, and it is thought that worm damage is the most likely cause of this form of colic. Treatment with painkillers and antispasmodic drugs relieves the pain, but may have to be repeated.

? WHAT IS FLATULENT COLIC?

Flatulent or 'gas' colic occurs when large amounts of gas are formed by fermentation in the stomach or intestines. This distends the bowels, causing pain which is usually continuous. This form of colic often follows overeating, or eating food material which ferments very easily (like grass mowings, wilted lucerne or clover). Food normally takes two hours to get right through to the large intestine from the stomach after eating, so flatulence develops soon after a feed. There may be increased bowel movement early on, and 'tinkling' and gurgling sounds may be heard. Later, bowel movement slows down and, if the large intestine is involved, the abdomen may become very distended. In very severe cases, pressure from the abdomen may cause difficulty in breathing, and the horse may even adopt a 'dog-sitting' position in an attempt to relieve this. Often, flatulence relieves itself by the passage of wind and acrid-smelling droppings from the rectum. Treatment involves the use of painkillers, and passing a stomach tube. The latter relieves gas from the stomach, and allows drugs which prevent fermentation (linseed oil and turpentine) to be given. Rarely, it may be necessary to perform exploratory surgery to relieve trapped portions of gas in the bowel, if painkillers and antifermenting treatment fails.

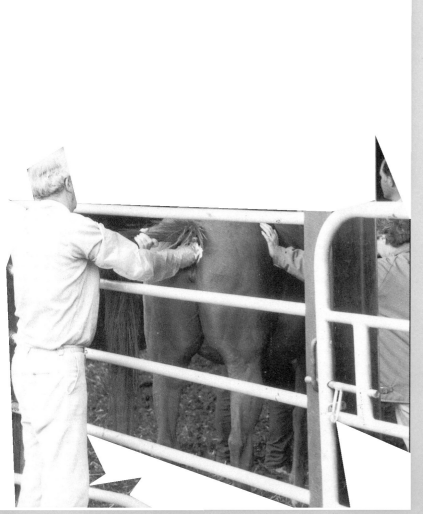

A vet giving a horse an **internal examination** to relieve possible blockage (*above*). Even though the horse has been tranquillized, the vet stands on the other side of a metal fence or gate, in case the horse should kick out.

disentangle the twisted portion of the bowel, must be undertaken as soon as possible. The final part of the small intestine is the section most commonly involved, and if bowel tissue has been strangulated for more than three or four hours, it may be necessary to remove the damaged portion altogether, and to join together the two 'ends' of healthy bowel on either side. During this process, very large volumes of electrolyte fluids — 30-40 litres (7-9 gallons) — must be given intravenously to combat the shock.

? WHAT IS THE COMMONEST CAUSE OF COLIC?

Worm damage is the commonest cause of colic. In a recent survey, it was estimated that current or previous worm damage was responsible for 90% of colic cases. Migrating redworm (*Strongylus vulgaris*) larvae damage the lining of blood vessels in the bowel. This blocks the vessels, depriving the bowel of blood and causing pain. Recently, several horses with problems around the junction of the small intestine,

A bran mash – bran mixed with warm water *(above)* – has virtually no nutritional value, but it acts as a mild laxative, so it is useful to regulate the horse's bowels.

? WHAT IS 'GRASS SICKNESS'?

'Grass sickness' is a serious form of colic, and one which has a most peculiar incidence. It occurs in certain parts of the United Kingdom only, and is most frequently encountered in mid-summer (May/June) although cases are reported at other times of the year. It usually only affects one animal in a group, and signs frequently develop after an individual has been moved to new premises (for example, a mare going to a stud to foal). Rarely, several animals on the same premises may develop the disease, one after another. In grass sickness the nervous control of bowel movement is affected, and the bowel becomes paralysed. Slowing and stopping of faeces in the terminal parts of the gut causes them to become impacted. At the same time, failure of the stomach to empty causes a mass of saliva and gastric contents to accumulate. When the stomach becomes over-distended, food and gastric contents are regurgitated through the nostrils – hence the term 'grass sickness'. The stomach may rupture, with fatal consequences. The initial signs of grass sickness are often indistinguishable from other forms of colic. In acute cases, however, the disease can progress very rapidly, and the horse may die of shock within 24 hours. More frequently, the process takes two to three days, and initial colic signs are followed by patchy sweating, twitching of skin muscles and shock signs. Diagnosis is confirmed by the absence of bowel sounds and the accumulation of fluids in the stomach. Euthanasia is frequently performed, to avoid further suffering. In mild cases, the onset of symptoms is much slower, and very slight bowel movement may persist. If animals with this lesser condition are fed gruel (soaked concentrate nuts) by stomach tube over a long period (of several weeks' duration), they may be able to survive, but are frequently permanently debilitated afterwards.

? WHAT IS IMPACTION, AND WHEN DOES IT OCCUR?

In the normal digestive process of a horse, large volumes of fluid are added to the food at the beginning of the digestive system – in the form of saliva and gastric juices. Much of this fluid is then reabsorbed (together with nutrients) in the intestines towards the end of the alimentary tract. If bowel movement – the muscular contractions that cause the contents to move along – is very slow, by the end of the bowel more fluid than normal has been absorbed, resulting in bowel contents which are very hard and difficult to pass on through the system. The horse is said to have a 'stoppage'. This situation frequently arises in stabled horses which may be receiving large amounts of dry food (hay, concentrates, and so forth). It is more likely to happen when horses drink insufficient water and have little exercise. To prevent the condition, laxative mashes are commonly given; salt (sodium chloride) and mild laxatives such as Epsom salts (magnesium sulphate) or Glauber's salts (sodium sulphate) are added to the diet to help stimulate horses' water intake. Signs are usually those of low-grade continuous pain (pawing the ground, looking at flanks, but not rolling), and the horse appears dull, disinterested and off its food. Treatment may involve painkillers, but will certainly include the administration of liquid paraffin and salts (Epsom, or Glauber's) by stomach tube, to help break up the impaction, and stimulate bowel movement. It is frequently necessary to repeat treatment by stomach tube, and it may take several days to clear the impaction.

One peculiar form of impaction occurs in newborn foals. Meconium is the mass of hard black faeces which is formed before birth; it is normally passed in the first few hours of life. If this does not happen, however, the foal may develop signs of discomfort and colic at any stage within the first two to three days after birth. This can usually be relieved by enemas of liquid paraffin or warm soapy water. If the problem persists, veterinary help should be sought. A special instrument is sometimes needed to remove hard material from further inside the rectum.

? IS THE CAUSE OF GRASS SICKNESS KNOWN?

In spite of intensive research over many years, the cause of grass sickness remains unknown. The disease appears not to be infectious. It may well be that it is a combination of factors in a way yet not understood that is involved in the development of the symptoms. Grass sickness can equally be found in stabled horses and its early signs are indistinguishable from other forms of colic.

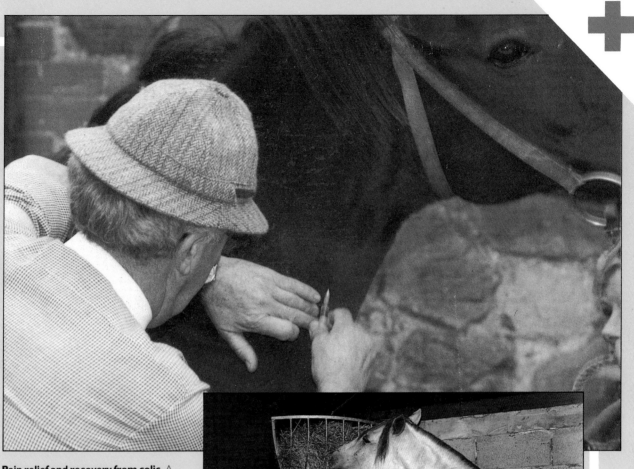

Pain relief and recovery from colic. A horse suffering from colic should always be encouraged to keep on its feet, and when it has quietened after the vet has given a pain-killing injection *(above)* it is always a good sign to see it eat some hay *(right)*.

 ### WHAT OTHER DIGESTIVE DISORDERS MIGHT MY HORSE SUFFER?

Worm parasites can cause enteritis and loose droppings (diarrhoea) in foals and adult horses. Foals may also 'scour' when their dam comes in season, presumably due to some physiological change in her milk. Acute enteritis can also develop in foals as a result of bacterial infection with organisms such as *Salmonella*. Acute enteritis less often develops in adult horses as a result of stress, or following the use of broad-spectrum antibiotics. The mechanism involved is not understood, and the condition is sometimes known as 'colitis X'. In cases where antibiotics are to blame, it seems that Clostridial bacteria have been able to multiply in the bowel following the antibiotic treatment, and the enteritis and shock is a result of the toxins they produce. In this condition, the horse becomes very depressed, loses its appetite, and rapidly becomes dehydrated and shows signs of shock. It may die within 12 hours; but more often, death follows diarrhoea of two to three days' duration. In milder cases, recovery can occur if intensive intravenous fluid therapy is successful in combating the electrolyte imbalance and fluid loss.

Loose droppings in adult horses can be a sign either of failure to reabsorb water in the bowel, or of enteritis. The former happens when the intestines become thickened by infiltration of the bowel wall by specific cells (mononuclear cells). This thickening affects the absorption of nutrients, and is usually permanent. The condition is sometimes known as 'malabsorption syndrome', and the bowel damage can be detected by blood tests. Although comparatively rare, it is one possibility that should be investigated .

RESPIRATORY DISORDERS

Respiratory problems such as virus infections, equine flu and strangles are quite common in horses. Guarding against the spread of infection and alleviating symptoms such as a cough are important measures,

? HOW ARE RESPIRATORY DISORDERS INVESTIGATED?

Other than lameness, respiratory diseases are the commonest cause of problems in horses. Although a great deal of useful information can be gained by listening to horses' lungs, and the causes of some infections can be detected by culturing nasal discharges, full investigation of respiratory problems was hampered in the past by the inability to examine the inside of the airways. This problem has been solved in recent years by the use of a fibre-optic endoscope ('scope) in equine veterinary practice. The instrument is a flexible tube composed of many thousands of fibre-optic strands, connected to a powerful light source; the light-carrying property of the strands enables an operator at one end to see through the other end. The tube is inserted through a horse's nostril into the nasal cavity, then pushed through the entrance to the windpipe (the larynx), and on down the windpipe itself to the lungs. By this means, the operator is able not only to get a clear view of the internal structures of the airways, and the way in which they are functioning, but — thanks to tiny attachments — to take samples (tissue or swabs) from any part. This instrument has revolutionized the investigation of airway problems in horses.

? MY HORSE HAS A 'RUNNY' NOSE: SHOULD I WORRY?

Horses suffer from several virus infections which cause flu-like symptoms. These usually produce a thin, watery nasal discharge. Thicker discharges are usually the result of bacterial infections (including

strangles) or infected sinuses. The most common respiratory virus that affects horses is Equine Herpes Virus Type 1 (EHV1). This causes mild flu-like symptoms which can include a watery nasal discharge, a slight temperature and swollen glands beneath the jaw; it may also cause coughing. The virus is easily spread from horse to horse, and is 'The Virus' that commonly affects racehorses and is responsible for loss of form. Immunity following infection is short-lived: recurrence is common. As yet, no effective vaccine is available. Treatment is not normally necessary, because recovery is straight-forward. However, it is important not to work horses during the course of the infection because it may prolong recovery.

Equine Influenza virus causes more severe flu-like symptoms. Affected horses are often quite ill, with a high fever. Glands under the jaw become enlarged, and the initial watery discharge often becomes thicker due to secondary bacterial infection. Horses invariably cough with this condition: coughing may be prolonged, and continue for several months after the initial infection has disappeared. This disease is highly infectious, and usually occurs in epidemics although it can easily be prevented by vaccination. Treatment of affected animals is usually with antibiotics to help reduce the temperature and to clear up the secondary bacterial infection; cough linctuses and electuaries may also be given to try to alleviate the cough.

? WHAT IS STRANGLES?

Strangles is a respiratory infection caused by the bacterial organism *Streptococcus equi.* It causes a thick nasal discharge, a fever sometimes accompanied by coughing, and enlargement of the lymph glands under the jaw. The incubation period is from two to six days, and the disease is highly infectious. It can be spread by direct contact between horses, or via tack, feeding or grooming kit. As the disease progresses the lymph glands under the jaw enlarge, causing the formation of abscesses which rupture, discharging pus through the skin. Abscesses may also be formed in other lymph glands beside the throat; more rarely, they form within the chest or abdominal cavity — this is a very serious complication, for such abscesses are hard to treat and sometimes prove fatal. Although antibiotic treatment may be helpful to alleviate fever in the early stages, its use may only temporarily suppress

the formation of abscesses: these tend to break out once the treatment stops. For this reason, vets sometimes withhold antibiotics and instead advise hot fomentation of jaw abscesses to help bring them to a head and clear them up more quickly.

Strangles is unfortunately becoming increasingly common, and any horse with suspicious signs (thick nasal discharge, large swollen glands and a cough) should be treated with due caution and isolated.

? MY HORSE HAS A NOSEBLEED: WHAT IS LIKELY TO HAVE CAUSED IT?

Blood from the nostrils is not common in horses. It may result from an injury to the nostril or to the side of the face, or come from a guttural pouch at the back of the mouth. In these cases the blood usually appears from one nostril only. Rarely, blood vessels in the sinuses at the base of the skull may haemorrhage into the nasal cavity, producing blood at both nostrils.

Racehorses sometimes bleed from the nostrils (about 1.5% of runners do this), and it was thought that the blood came from the nasal cavity. However, use of the endoscope and X-ray examination have shown that this blood originates from a particular part of the lungs. Furthermore, it was discovered — to many people's surprise — that a high percentage of racehorses (60-70%) bled from the lungs during exercise, and that this blood passed up the windpipe and was swallowed, and only in extreme cases appeared at the nostrils. Later investigation has shown that many other types of horse have lung haemorrhage at exercise, including eventers, show-jumpers, and even ponies competing in gymkhanas. As yet no satisfactory explanation is available for this finding, but it is so widespread that some physiological factor must be involved. A wide variety of possible causes has been put forward. Lung haemorrhage must affect performance, and it would seem that so-called 'ungenuine' racehorses may, in fact, be avoiding over-exerting themselves in order to prevent lung haemorrhage.

Sponging the nostrils with a clean sponge *(right)* is part of the everyday grooming of a horse. However, if a horse has a runny nose caused by a cold or 'flu, it should be gently sponged several times a day. Wash the sponge thoroughly after each use, and do not use it on any other horse.

DO HORSES SUFFER FROM SINUS INFECTIONS?

Horses have several large air-spaces within the bones of their skull, which help to reduce the weight of the head. These include the large frontal sinuses beneath the forehead, and the maxillary sinuses below the eye. The sinuses on either side of the head do not communicate with each other, but drain directly into the nasal cavity on each side. Sinuses easily become infected with bacteria from the nasal cavity, resulting in a thick discharge from one nostril. This condition frequently responds well to antibiotic treatment. Measures such as feeding on the ground, and even steam inhalation, may also help infected sinuses to drain. If the nasal opening of the sinus becomes blocked, pus accumulates within the sinus. In this case, it may be necessary to perform surgery to establish drainage. This can be achieved either via the nasal cavity or through holes in the bones of the side of the face. Infection of the maxillary sinus may result from problems (abscesses) at the roots of the upper cheek teeth.

WHAT ARE GUTTURAL POUCHES?

Guttural pouches are two small cul-de-sacs (off the Eustachian tubes) on either side at the back of the mouth; they lie under the muscles beneath the ear. Among domestic animals they are found only in horses. They appear not to have any useful function, and the reason they have evolved is unknown. On the roofs of these pouches run important blood vessels and nerves. Infection, especially by fungi from the nasal cavity, can sometimes enter these pouches, and they can fill with air (especially in young foals) or pus. Such infection may damage the blood vessels (causing haemorrhage) or the nerves (causing problems in swallowing). The guttural pouches can be examined with an endoscope via the nostrils, and tubes can be inserted into them to permit irrigation, drainage or administration of medication.

WHAT IS A ROARER?

For many centuries a condition in horses has been known in which affected animals make a characteristic inspiratory breathing noise during exercise. Horses with this problem were known as 'roarers' or 'whistlers', depending on whether the sound they made was low or high pitched. It was thought to be a hereditary condition, and such animals were considered 'unsound'.

The main symptom of the condition is wasting of the muscles around the entrance to the windpipe (the larynx), particularly on the left side. These muscles are paralysed due to damage of the nerve which supplies them. The condition is thus now also known as laryngeal paralysis, or, because usually only one side of the larynx (the left) is involved, it is also known as laryngeal hemiplegia. The reason for the nerve damage is unknown.

Roaring more commonly affects larger types of horses, and is not found in ponies. The characteristic inspiratory breathing noise made at exercise can often be heard on television coming from some of the larger show-jumpers during jumping competitions! Reduced air-flow to the lungs causes a loss of performance, but because the paralysis is usually not complete, how much the performance is affected depends on the degree of obstruction.

Provisional diagnosis can be made by listening to a horse's wind at exercise. This is usually checked using an endoscope, by which means the cartilage and vocal cord on the left side of the larynx can be seen 'flopping' into the middle of the airway, in horses suffering from paralysis. However, some horses which show signs of laryngeal paralysis using an endoscope make no noise at exercise, whereas a similar noise can sometimes be produced by horses with other problems (such as lymphoid hyperplasia of the pharynx — the equine equivalent of tonsillitis), and not showing paralysis. It would seem that performance is most likely to be reduced in horses showing paralysis *and* making an abnormal inspiratory noise at exercise.

CAN A ROARER BE TREATED?

There is nothing that can be done to restore function to the damaged nerves or wasted muscles. Treatment of roarers involves trying to improve the airflow to the lungs. The most common surgery performed is the so-called Hobday operation: an incision is made into the larynx, and the lining of two small saccules lying behind the vocal cords is removed; when this wound heals, adhesions develop which pull the obstructing laryngeal cartilage, and vocal cord, to the side of the airway. More refined operations have also been devised, using an elastic implant to hold the cartilage to the side of the airway. None of these operations is completely successful, but they can improve the airflow, and thus restore performance. In the past, the problem was bypassed by inserting a tube into the windpipe, through which the horse could breathe. This has several disadvantages, and 'tubing' is seldom performed nowadays.

WHAT IS C O P D?

Many stabled horses and ponies develop a harsh, dry cough, but otherwise appear well. In the past, this was known as 'broken wind' and was thought to be caused by straining the lungs through over-exercise. Severely affected animals often have very laboured breathing, and the disease was sometimes known as 'heaves'. More recently it has been discovered that the condition is due to lung damage (emphysema) resulting from an allergic reaction to the spores of moulds present in the stable atmosphere, from fodder and bedding. When these spores enter the lungs, they stimulate cells (mast cells) in lung tissue to release chemicals which cause constriction of the smaller airways (bronchioles); this results in obstruction of the airflow within the lungs. The disease has become known as Chronic Obstructive Pulmonary Disease, or COPD.

Signs of COPD do not occur in horses at grass. When a horse which is sensitised to mould spores is stabled, however, it develops a harsh, dry cough which is often heard at the beginning of exercise but frequently disappears as the horse warms up. Normal expiration in horses is a passive process, relaxation of the chest following without effort after its expansion during inspiration. But in animals suffering from COPD, an extra, active effort is required to empty the lungs through obstructed airways, necessitating the voluntary contraction of the chest muscles following their relaxation during expiration. This can usually be observed by watching the movement of the horse's ribcage at rest. Affected animals show a 'double-lift' of the ribcage in expiration. In severe cases, this extra effort develops chest muscles to such a degree that a so-called 'heaves' line can be appreciated. In very severe cases, horses can be in great distress and have extreme difficulty in breathing: they may wheeze, and can have very blue visible mucous membranes. This can be an extremely serious problem in mares during late pregnancy.

CAN PREVENTIVE MEASURES BE TAKEN?

In most instances the problem can be lessened, or removed, by preventive measures aimed at reducing spore levels in the stable environment. Research has shown that nearly all fungal spores are of such a small size that they can be inhaled into the lungs — they are not filtered in the nostrils. Hay and straw produce large amounts of fungal spores, particularly when they are 'fausty'. Deep litter bedding, with other materials, also produces high levels of spores. Research has shown that the highest levels are produced when straw bedding is shaken up, so it is advised that no horse should be in the stable while this is taking place. In fact, horses should preferably not be returned to the box for at least 20 minutes, in order to allow the spores to subside. Ventilation is vitally important in the build-up of dangerous levels of spores: good-quality hay and straw may produce dangerous levels if ventilation is inadequate. Feeding hay on the ground (and not from a hay rack above head level) reduces spore levels, and is preferable to using a haynet. However, soaking hay (in a net) in water for five minutes reduces spores to a minimum.

Recently, drugs similar to those used to treat asthma in humans have become available for horses. These are administered via a face-mask connected to a nebulizer (a sort of atomizer that produces a fine mist of the drug with the help of an air compressor). Treatments are given for 20 to 30 minutes daily over four consecutive days; one such course prevents symptoms for a period of three weeks. The drug stops mast cells releasing the chemicals responsible for constriction of the airways. Like turning affected horses out to grass, this is not a cure, but it is effective in preventing symptoms. Mould allergy is widespread in the equine population and a serious cause of ill-health. It is particularly important to avoid exposing young horses to high levels of mould spores and to the possibility of sensitising them.

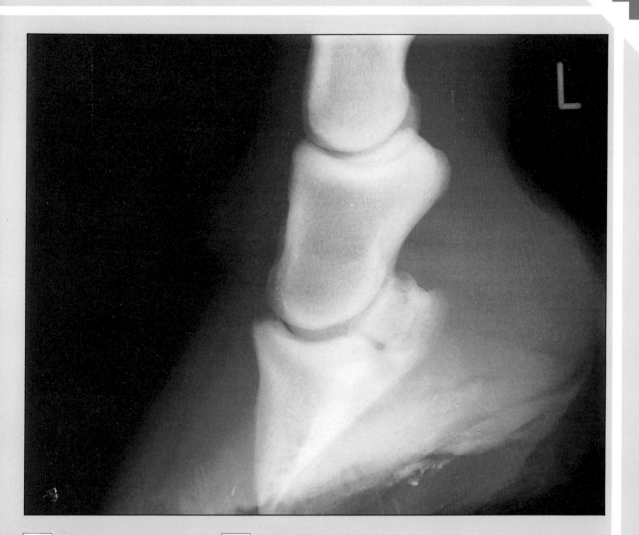

The results of severe **laminitis** – inflammation of the sensitive laminae under the wall of the foot *(above)*.

? MY HORSE IS COUGHING: WHAT IS THE MOST LIKELY CAUSE?

If the horse has a moist cough, a runny nose, and other animals are also coughing, it is most likely that coughing is a result of a virus respiratory infection or strangles. If the horse is stabled, appears well, and has a harsh, dry cough with no enlarged lymph glands, it is most likely that it is suffering from COPD. If the horse is at grass, or has in the past shared grazing with a donkey, it may well have picked up lungworm infection. Coughing from virus diseases or secondary bacterial infections responds to antibiotics and cough medicines, whereas coughing due to COPD or lungworm do not. The latter can be treated by drugs (by inhalation or worming) or preferably can be prevented by avoiding high spore levels in stables, and not sharing grazing with donkeys.

? WHAT OTHER LUNG PROBLEMS MIGHT AFFECT MY HORSE?

Pneumonia may result from bacterial infection either brought on by stress (such as severe chilling) or spread from infection elsewhere in the body. Not common in older horses, pneumonia is more common in foals, which often develop a harsh cough and suffer considerable weight loss as a result. In the foals, infection with *Corynebacterium equi* is serious, because this organism tends to cause large abscesses to form in the lungs, which antibiotics are unable to penetrate. Pleurisy (inflammation of the inner lining of the chest wall and outer covering of the lungs) is a complication that is sometimes encountered in horses with pneumonia.

Lung cancer is very rare in horses. It may be seen in old age, when it has spread from tumours elsewhere in the body.

FOOT PROBLEMS

Detecting lameness is not always easy for the novice. Having determined which leg is lame, it is worthwhile checking something simple is not causing the problem. A vet will then carry out further treatment.

HOW CAN I TELL IF MY HORSE IS LAME?

Detecting lameness is not always easy, especially for the novice. In many instances, lameness is very slight or even transient, and the animal may warm up during exercise or, conversely, show signs only after very hard work. Often it is the rider who first notices that something is wrong, that the horse is not striding out properly. It may feel 'unbalanced', particularly when turning or carrying out precise manoeuvres in dressage. Although lameness can sometimes be obvious at the walk or the gallop, the trot is the best pace for its detection and investigation. When trotting, a lame horse tries to avoid pain by reducing the weight put on the lame limb, and it does this by altering the position of its head or hindquarters.

HOW CAN I SPOT WHICH LEG IS LAME?

To detect lameness, ask someone else to trot the horse up for you on a hard level surface, on a loose rein (at least 30 centimetres — one foot — of rein, or leading rein). For foreleg lameness, the horse should be trotted towards you; the horse is lame if it is seen to nod its head — the head goes up when the lame foreleg hits the ground. Hind leg lameness is easier to spot when the horse is trotting away from you. In this case, watch the hindquarters: the quarters on that side are raised when a lame leg hits the ground — again in an attempt to avoid putting weight on the lame limb. It is essential to watch the movement of the head and hindquarters, and to relate these to footfall. It is easier to remember SINKS ON THE SOUND SIDE — for the

LAMENESS

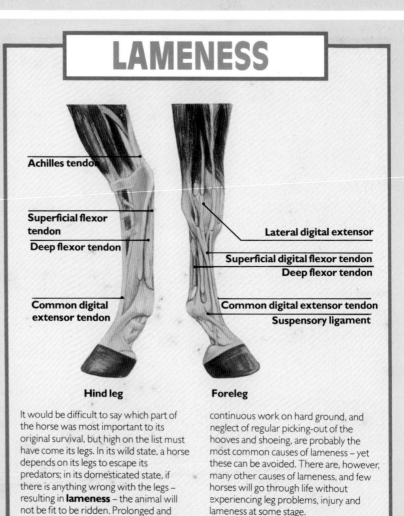

Achilles tendon

Superficial flexor tendon

Deep flexor tendon

Common digital extensor tendon

Lateral digital extensor

Superficial digital flexor tendon

Deep flexor tendon

Common digital extensor tendon

Suspensory ligament

Hind leg

Foreleg

It would be difficult to say which part of the horse was most important to its original survival, but high on the list must have come its legs. In its wild state, a horse depends on its legs to escape its predators; in its domesticated state, if there is anything wrong with the legs – resulting in **lameness** – the animal will not be fit to be ridden. Prolonged and continuous work on hard ground, and neglect of regular picking-out of the hooves and shoeing, are probably the most common causes of lameness – yet these can be avoided. There are, however, many other causes of lameness, and few horses will go through life without experiencing leg problems, injury and lameness at some stage.

head and hindquarters, in foreleg and hind leg lameness respectively. Unfortunately, at the trot the head also tends to drop slightly when a lame hind limb hits the ground, and confusion could thus arise. Care must be taken not to mistake a hind leg lameness for one in the diagonally opposite foreleg — in foreleg lameness, the hindquarters remain level and unaltered. When a horse is lame in both legs, there is usually no head or quarter movement to help. This can be more difficult to detect, although these horses usually have a shuffling, 'pottery', gait in front, or take very shortened strides behind. There is often a thin borderline between poor action and lameness. In dressage horses with back or muscle injuries, detection of a problem is frequently a matter of watching the way the feet move in flight, and analysing the character and length of the stride. The movement of different limbs is then compared when the horse is lunged, or ridden in a circle, in opposite directions.

WHAT CAN I DO TO HELP PINPOINT THE TROUBLE?

Having determined which leg is lame by watching the animal trotting, it is always worthwhile examining further to ensure that something simple is not causing the problem (such as a thorn, a cut, or a stone under the shoe). Heat, swelling and pain usually indicate the site of the problem. In the foot, swelling is impossible, but heat is nearly always associated with foot problems. By putting a hand on the hoof wall, and comparing it with the opposite hoof, it is usually possible to detect differences in temperature. This is easy when infection is present (pus in the foot, or pricked foot following shoeing), or in laminitis. It should

Looking for **foreleg lameness** – the horse is trotted towards the observer on a firm, level surface, making sure his head has at least 15cm (6 ins) of free rein *(above)*. If lameness is present in a front leg, the horse will nod its head, raising it when the lame leg lands on the ground.

Underside view of the same foot *(above)*, showing that the animal is suffering from a condition known as **seedy toe**. This is an inflammation of the coronet which makes the horn of the hoof separate from the foot. When the hoof is tapped it will sound hollow; lameness always results from this condition.

not normally be easy to detect a pulse in the digital arteries. These run on either side of the back of the pastern, at the side of the deep digital flexor tendon. In laminitis, or with pus in the foot, a rapid, throbbing pulse can usually be felt in these arteries. Having established that the foot is cold and is being put to the ground normally, the rest of the affected limb can be felt to look for signs of heat and swelling. In the foreleg, particular attention should be paid to the tendons that run down the back of the cannon bone. Any enlargement in the two flexor tendons, or in the check and suspensory ligaments (which lie between the flexor tendons and the cannon bone — see diagram) should be viewed with suspicion. Likewise, swelling and heat around the fetlock joint is likely to be a sign of arthritis, and this is more common in the front limbs. Splints can be a cause of soreness in the forelegs of young horses. In the hind limbs particular attention should be paid to the hock joint — look for heat and swelling in and around this structure.

A vet normally uses special hoof pincers to apply pressure to various parts of the sole to test for foot problems. In addition to a thorough examination of the limbs, he will probably lift them up and flex the various joints to check for pain and other evidence of arthritis. Some lamenesses (particularly hock problems) show up better if the affected limb is held up (to flex the limb) before trotting the horse up. He may also back the horse, or turn it in a tight circle, to help detect some conditions (back and hip problems). If the site of lameness is still not apparent, nerve blocks may be performed. This involves injecting local anaesthetic around the nerves at various parts of the limbs. This effectively removes sensation from the limb below the injection site. By this means, the horse is made 'sound' if the problem lies below the injection point. Injections at the back of the pastern thus 'block' the foot; those below the knee 'block' the fetlock and foot; and injections in the forearm and above the hock 'block' the knee and hock joints and all the lower limb below them. In some instances, it is not until the lame leg is made 'sound' by a nerve block that it can be appreciated that a problem exists in the opposite limb. Local anaesthetic may also be injected into joints, in order to make certain that this is the site of injury. This can be done as a check before expensive drugs are injected into joints to treat degenerative joint disease (arthritis).

X-ray examination plays an important role in lameness examination, particularly in investigating conditions inside the hoof. Although good radiographs of limb bones can be obtained from the elbow and hock joints and below using portable X-ray equipment in the stable, examination of the shoulder, stifle, hips and back requires longer exposure and more powerful equipment. This means that the horse will almost certainly have to be anaesthetised, and be referred to a specialist establishment that possesses such expensive equipment (university veterinary schools, or the Animal Health Trust at Newmarket). More recently, ultra-sound scanners have been used to investigate tendon injuries in horses.

MY HORSE HAD A HARD DAY YESTERDAY; TODAY IT IS LAME AND HAS A SWELLING AT THE BACK OF THE CANNON BONE – HAS IT SPRAINED A TENDON?

Tendon injuries are one of the commonest and most troublesome causes of lameness in performance horses. The superficial digital flexor tendon (SDFT), which runs from the muscles of the forearm down the back of the leg immediately below the skin to the pastern bone, and the deep digital flexor tendon (DDFT), which lies immediately beneath (in front of) the SDFT and inserts on the pedal bone (within the hoof), are most important structures. Both these tendons are stressed repeatedly at fast paces and may be sprained when the horse becomes tired. Initial injury is usually slight, with a small amount of haemorrhage in the sheath surrounding the tendon. A few of the small tendon fibres are also damaged, and there may be slight heat or swelling which can be felt through the skin of the foreleg. If the horse is not rested, or the damage is more serious, more fibres are damaged, with larger amounts of haemorrhage and a correspondingly greater distension of the tendon. The ability of tissues to heal depends upon their blood supply, and, because tendons and ligaments have a very poor blood supply, their repair is very slow. Also, when tendon tissue heals, scar tissue is formed rather than new tendon tissue. Scar tissue does not have the strength of tendon, and is likely to suffer re-injury. Thus, once a tendon is damaged, injury is likely to recur, particularly if the horse is not rested. In extreme cases, this can even result in complete rupture of the tendon. The SDFT is more frequently sprained, but sprain of the DDFT can occur on its own, or, more commonly, accompany a sprain of the SDFT. Damage in the DDFT is always more serious.

WHAT DOES TENDON TREATMENT INVOLVE?

Treatment for recent tendon sprains involves cold hosing or ice-packs to reduce the inflammatory reaction. Supporting bandages on both forelegs are essential (see the First Aid section). Long periods of rest are required – it takes six months for tendons, and nine months for ligaments, to repair. Many different procedures have been tried to strengthen and improve the repair of tendon injuries. In the past, firing with a hot iron and blistering were used to try to increase the blood supply to the tendon, and thereby improve healing. These treatments applied to the skin (particularly line firing) have been shown to have little or no effect on the blood supply to the tendon, and are now not generally recommended. Their apparent beneficial effects may have only been achieved by enforced rest. Various surgical operations have been devised to try and improve the blood supply to damaged flexor tendons. These involve making an incision into the tendon – commonly known as 'tendon splitting'. Recently it has been discovered that if strands of carbon fibre are implanted into damaged tendon, new tendon tissue forms along the direction of the carbon fibres. Because the new tendon tissue that is formed appears to be stronger and less liable to re-injury than the scar tissue which would otherwise result, carbon fibre implants would seem to have a distinct advantage over previous surgical methods of treating sprained tendons. Many other forms of therapy have been used to treat tendons, including ultra-sound, electromagnetic and laser therapy. Tension on the tendon is sometimes relieved by the temporary fitting of a shoe with a raised heel. Plaster casts are occasionally fitted around the limb, to give support. Although the clinical signs of sprained tendons are usually obvious, increasing efforts have recently been devoted to trying to detect very early changes in the tendon. Techniques which are being used for this purpose include the use of force plates (computer analysis of the way in which a horse places its feet on a force measuring plate on the ground), highly sensitive equipment for detecting small changes in temperature (thermography), and ultra-sound scanners.

WHAT IS THE CHECK LIGAMENT?

The check ligament is a small ligament running from the back of the knee in a horse's foreleg, down to the middle of the back of the cannon bone, where it inserts on to the suspensory ligament. The check ligament sometimes becomes sprained as a result of a severe injury to the limb, and swelling occurs about a hand's breadth below the knee, at the back of the cannon bone. Because the check ligament lies so deep in the leg, it is very hard to treat, and the injury tends to recur. All too often this is a result of insufficient rest because, as ligaments have a very poor blood supply, they take at least nine to ten months to repair, and a minimum of 12 months' rest is necessary. As with strains of the suspensory ligament, permanent thickening may occur, and the animal's work may have to be restricted to lessen the chance of re-injury.

WHAT ARE SPLINTS?

Small, thin, elongated splint bones are present on the inside and outside of the cannon bones on all four limbs. These are the small remains of bones that once were connected to the second and fourth digits, which have been lost during evolution. The splint bones are joined to the cannon bone by strong ligaments. Young horses frequently develop a hard swelling over the upper end of the inner splint bone of one, or both, forelegs. This is the result of concussion on immature bone, which causes an inflammation in the outermost layer of the bone (periostitis). Soreness and swelling develop just below the knee, and the horse may be lame or have a stiff action. New bone is formed as a result of the inflammation, and quite a large, hard lump can result. Treatment involves rest and cold hosing or ice-packs, in the initial stages; this can be followed by topical (local, surface) treatment with a liquid containing cortisone combined with another drug that helps it to be absorbed through the skin, to further reduce the inflammatory swelling. A mild 'working' blister may also be used after 10-14 days to stimulate increased blood supply and reduce the swelling. This condition must be distinguished from fractures of the splint bone, which can result from kicks and blows, and which cause a larger callus. A fracture can of course be detected on X-ray; these usually heal with rest alone. Rarely, it may be necessary to remove a fractured piece at the bottom of a splint bone, if it does not heal, because it may otherwise rub against the flexor tendons.

MY HORSE HAS A SWOLLEN KNEE BUT IS NOT LAME: SHOULD I WORRY?

A large tendon sheath runs down the foreleg, across the front of the knee. This is frequently knocked by horses who fail to jump obstacles. If this happens repeatedly, reaction and a large swelling (hygroma) can set in over the front of the knee joint. This does not make the horse lame, but is very likely to enlarge even more if knocked again. The swelling is very slow to respond to treatment, and rest is essential; it may be necessary to drain the fluid contents of the swelling, which is sometimes found to include blood from a ruptured vessel. The knee must be bandaged after drainage. Permanent thickening frequently results from such an injury, although blisters have sometimes been used in an attempt to reduce this. A similar permanent swelling occurs as a result of grazes to the knee (from going down on a road): this is sometimes called 'broken knees'.

These conditions, which do not cause lameness, must be distinguished from problems within the knee joint, which can cause lameness. Inflammation of the knee joint (carpitis), or small (chip) or larger (slab) fractures of the small bones within the knee joint, are uncommon except in young racehorses, but may produce marked distension of the knee joint, and can be detected by X-rays.

? WHAT ARE WIND GALLS, AND ARE THEY SERIOUS?

Wind galls or wind puffs are a horseman's terms for soft swellings on either side of the legs at the fetlock joint. Swellings at the back of the leg are due to distension of the tendon sheath. Known as tendinous wind galls, they do not cause lameness, and are a blemish that can often be seen in horses that are accustomed to hard work. No treatment is necessary. But rest, and supporting bandages, can help reduce the swelling when it first appears. It is important to distinguish this swelling in the tendon sheath (which lies behind the suspensory ligament) from distension of the fetlock joint capsule (which occurs between the suspensory ligament and the cannon bone) which is known as an articular wind gall. These swellings of the joint capsule are signs of arthritis (degenerative joint disease) within the fetlock joint, and may be associated with lameness. Articular wind galls are a sign of wear and tear, and possible joint problems. They do require treatment. Wind galls must be distinguished from swelling under the skin around the fetlock joint, which can result from poor circulation or insufficient exercise.

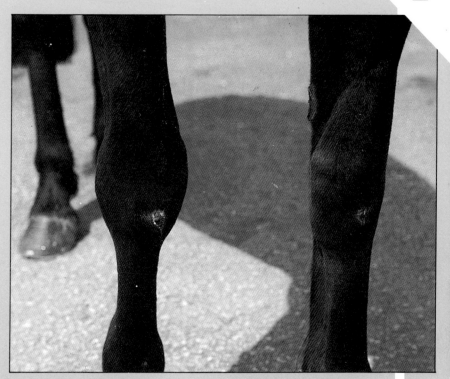

Windgalls. Soft, puffy swellings, which appear just above and behind the fetlock joint on hind and forelegs *(below)*. These are the result of strain and concussion, although they rarely cause lameness. However, they are unsightly, and are a sign of bad management. The horse should be rested and given only gentle exercise until the swelling has gone.

Broken knees. Scabs on the knees show that the horse has tripped and stumbled, probably on stony ground *(above)*. This could be the result of neglecting the feet – letting them grow too long – or it could be through careless riding. The hairs will grow through white, as a permanent reminder of the injury. This evidence of broken knees could detract from a horse's worth.

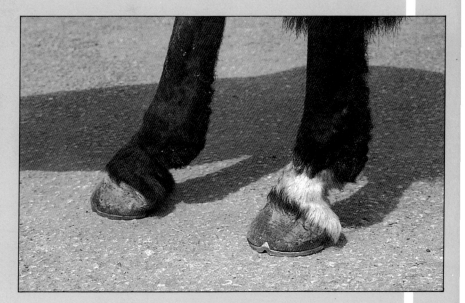

? WHAT IS DEGENERATIVE JOINT DISEASE?

This is a form of inflammation within joints (arthritis) resulting from the repeated effects of concussion (wear and tear). Changes occur in the composition of the fluid within joints: it becomes thickened, and is less able to lubricate the joints effectively. At the same time, chemicals released into the joint as a result of the inflammation damage and destroy the joint cartilage and underlying bone. In attempting to heal, new bone is formed, particularly around the edges of the joint, which may provoke further reaction when pieces of new bone become dislodged during movement. Although many joints can be affected in this way, the commonest sites of trouble are the foreleg fetlock joints. Affected animals have hot, swollen joints. The swelling is noticeable at the front and at the sides of the fetlock joint (articular wind galls), and in severely affected animals makes them look as if they have 'apple-like' fetlock joints. It is often difficult to detect the bone changes on X-ray, but recently an instrument for looking into joints (an arthroscope) has been more widely used in equine veterinary practice. Through this, erosions of the joint surface can easily be seen. In the past, such animals were treated with anti-inflammatory and painkilling drugs like phenylbutazone ('bute') orally. Such drugs have side effects, and affected the whole horse rather than just the joint. It is now considered more desirable to treat the joint, or joints, involved directly, by injecting drugs into them. Injections containing hyaluronic acid have been found to simulate the properties of 'joint oil', to improve lubrication, and to prevent further damage. A new drug (polysulphated glycosaminoglycan) has been found actively to stimulate repair of damaged joint cartilage. Treatment with these drugs may require one or more injections into affected joints by a vet. Because they are expensive, it is often helpful to 'block' the joint by injecting local anaesthetic into it, to ensure first that it is the correct, and only, cause of lameness. Rest (to allow the joint time to heal) and cold hosing, ice-packs and physiotherapy (to reduce the inflammation and speed up healing) are also helpful. In advanced cases, chronic lameness may result, requiring continual use of low doses of painkillers (like 'bute'), or retirement from work.

? MY HORSE HAS FORELEG LAMENESS, AND I CAN'T FIND ANYTHING WRONG IN THE LOWER LIMB – COULD THE TROUBLE BE IN THE SHOULDER?

Shoulder lameness is rare in horses, except as a result of kicks, or falls when jumping. In a recent survey, many horses suspected of having shoulder lameness were found to have problems further down the limb. This was confirmed by nerve blocks. In shoulder lameness, the leg is held very stiffly, and is not flexed or moved forward when the horse moves. The head is raised very high when any weight is put on the lame leg. In severe cases the horse may stand resting the affected leg on the toe behind the opposite foreleg.

A condition of bone degeneration, which may also be found in the elbow, hip, stifle and hock, is sometimes found in the shoulder joint of young horses. This condition (osteochondritis dissecans) is thought to be associated with a lack of blood supply to the joint cartilage.

? WHAT IS RINGBONE?

Ringbone is a horseman's term for new bone which is formed around the pastern bones as a result of injury (a blow, or repeated concussion), or of arthritis. The new bone formed can usually be detected on X-ray examination, and may cause heat and swelling, which can be felt. The effects of ringbone depend entirely on its position. New bone formed within the pastern or coffin joints (articular ringbone) is very serious, and frequently causes permanent lameness. Ringbone outside the joint is due to damage to the outer layer of the bone (periosteum) and may, in time, settle down and cause no further problems.

Ringbone is sometimes classified as high or low ringbone (depending on whether it is on the long or short pastern bones), articular (within the joint), or extra-articular (outside a joint). Treatment involves rest and the use of anti-inflammatory drugs in the initial stages. Blistering may help new bone outside joints to 'settle down'. Injections into the pastern joint are not very successful in relieving this problem, which is more common in forelegs than hind legs. In particularly valuable horses used for competition, complicated surgery to permanently join the bones of the pastern joint together (arthrodesis) has sometimes been performed with success.

? MY HORSE'S HIND LEG 'LOCKS' BEHIND IT: WHAT COULD BE CAUSING THIS?

Horses are sometimes discovered in a stable with one hind leg sticking out behind the body. The horse is unable to flex its legs, nor can the leg be picked up by the owner. This is more common in foals and yearlings, particularly if they are weak and have a very upright (straight) hind limb conformation. It may also be found in older horses that are suddenly taken out of training and lose muscle quickly. The condition is caused by the inner ligament of the three ligaments joining the horse's patella (knee-cap) to the bone below the stifle joint (equine knee) becoming locked over a notch at the lower end of the thighbone (femur). This prevents the horse flexing both its stifle and hock. Horses stuck in this manner can often be relieved by making them back (walk backwards). If this fails, manual pressure exerted on the outside of the stifle, attempting to push the patella inwards and downwards, may release it. Lesser degrees of this problem are commmon, and the hind leg may 'catch' when the horse walks, rather than locking. Both legs are usually affected, and the problem is likely to recur. Whereas young horses may grow out of it, older animals often require surgery to overcome the problem. This involves cutting the inner patella ligament (usually of both legs). After a period of rest to allow the horse's action to return to normal, no further problems should arise.

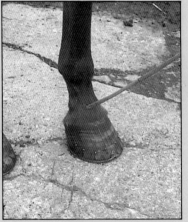

A clear case of **ringbone** – a bony formation which occurs in the pastern region. It is sometimes caused by too much work on hard ground, but it is also a hereditary complaint. It generally results in chronic lameness.

MY HORSE HAS A SWOLLEN HOCK: WHAT COULD HAVE CAUSED THIS?

The hock is the hardest-working joint in the horse's body, and is responsible for absorbing much of the concussion in the hind limb. As a result, it is a common site of strain and injury. A 'thorough-pin' is a swelling occurring in the tendon sheath above the hock joint, causing a distension either side of the Achilles tendon where it inserts on to the bone at the back of the hock. This swelling is a sign of overwork, especially in young, immature horses. It does not cause lameness, and may reduce with rest. If it does not subside, the swelling can be drained by a vet. The swelling of a thorough-pin must not be confused with distension of the joint capsule of the hock joint, which is known to horsemen as a 'bog spavin'. This swelling also occurs on either side of the hock joint, but lower down than a thorough-pin. At the same time, the front of the hock joint is distended, and if pressure is applied here, increased tension in the swellings at the back of the hock can be appreciated (this does not occur with a thorough-pin). Bog spavins are frequently associated with poor conformation (straight hind legs), which put an extra strain on this joint. Sudden sharp movement, as occurs in activities such as polo and show-jumping, may strain the joint capsule, causing bog spavin. In most cases, no heat, pain or lameness results. In the early phases, cold hosing, rest and application of a pressure bandage may be recommended. If the condition becomes chronic, veterinary treatments ranging from blistering to draining and injection of anti-inflammatory drugs into the joint may be used.

A third type of swelling occurs around the hock joint. This is a hard, bony swelling on the inner lower aspect of the joint (just above the top of the cannon bone) which is known by horsemen as a bone spavin. This is caused by arthritis in the small bones at the bottom of the hock joint. Poor conformation ('cow' and 'sickle' hocks) is often associated with bone spavin. The problem can also be caused by sudden strain on the joint (polo, jumping and similar activities). The signs of this condition are variable, but affected animals do not pick their hind feet up, take short strides, and may wear the toe of their hind shoes. The degree of lameness is not related to the size of swelling. Horses with spavin can be very lame and have little or no swelling. Sometimes the arthritis 'settles down', and horses can have large, hard swellings with no

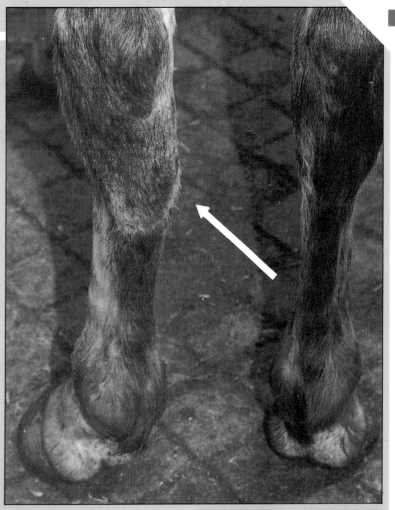

Swelling just above the fetlock joint of a hind leg *(above)* indicates **ligament strain**.

lameness. Lameness may be produced by holding the hind leg up (to flex the hock) for one minute, before trotting the horse away (a so-called 'spavin test'). The position and extent of new bone formation can be detected on X-rays, and injecting local anaesthetic around the 'seat of spavin' can help confirm that this is the cause of lameness. Treatment may include rest, combined with corrective trimming and shoeing (raised heels and rolled toe). Anti-inflammatory drugs can be helpful, and pin firing is sometimes used to try and reduce chronic bone enlargement. Alternatively, surgery may be performed to remove a section of the tendon that runs over the 'seat of spavin', to ease the pain which can result from pressure of the new bone on this tendon. Spavin is frequently a recurring problem, and buying a horse with this 'unsoundness' should be avoided.

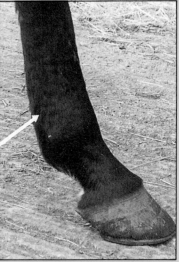

A case of **tendon strain**, seen by the 'bowed' outline of the area arrowed *(above)*. This condition, which results in swelling and lameness, is brought about by too much work on hard ground.

WHAT ARE CURBS?

A curb is a horseman's term for a swelling on the back of the hind leg, about a hand's breadth below the point of the hock. It is found in young horses, especially those with weak hocks — 'sickle hocks', in which the angle of the hocks is very curved when viewed from the side (a conformation that is thus also known as 'curby hocks'). The swelling is due to the strain of a ligament (plantar ligament) below the hock. This results from sudden pressure on the back of the hock from jumping or pulling heavy loads. It is often a result of overworking young horses. Both legs are usually affected, although one is frequently much worse than the other. There is soreness, heat and swelling at the site of curb. The animal can be lame at first, and may rest the leg by raising the heel. Treatment in the early stages involves rest, cold hosing or ice-packs; liquids containing cortisone combined with an agent that helps it to be absorbed through the skin can be painted over the area to reduce the inflammation. In the later stages of healing, a mild blister (a working blister) is sometimes applied daily over the area to try to stimulate healing of the ligament. Raised heel shoes may also be fitted. Although there may be signs of soreness and lameness when curbs first develop, this is not always the case, and some horses show no lameness at all. Sometimes new bone is formed at the insertion of the ligaments, making a hard permanent swelling. In most instances, scar tissue formed as a result of the strain leaves a thickening, resulting in a permanent blemish. This does not usually cause trouble, however, and if young horses that are weak are given more time to develop, they should have no further problems.

WHAT IS JOINT-ILL?

Joint-ill is the commonest form of lameness in young foals. Often, this condition is not appreciated at first because owners think that the animal has been kicked or injured, because lameness may occur before any heat or swelling can be detected in a joint.

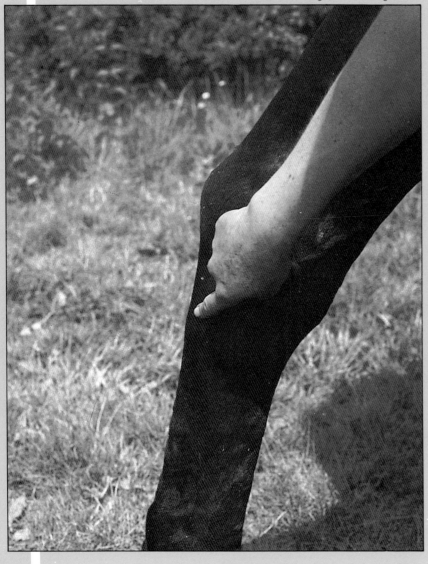

Hosing the leg with cold water *(above)* for ten to fifteen minutes at a time will help to reduce the swelling.

Swelling and thickening of the tendon or ligament just below the hock, as arrowed *(Left)*, is evidence of a **curb**. This is caused by strain and concussion;

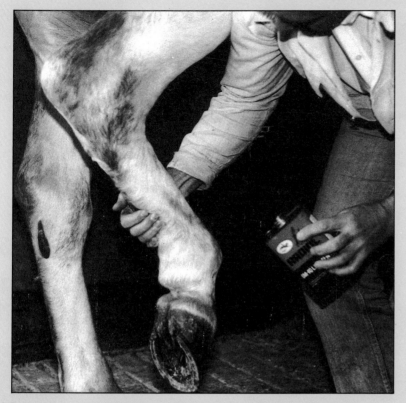

The leg should be dried off after each hosing *(above)*, then rubbed with a **liniment oil**.

The joints of newborn animals have a particular attraction for bacterial organisms from elsewhere in the body. These frequently enter via the umbilical chord (navel) soon after birth. Any joint in the body may become infected, but stifle joints are the most frequent source of trouble. In addition to lameness, heat and swelling can usually be felt, and the foal has a temperature. It may appear 'sleepy' and refuse to suck from its dam. It is essential that large doses of antibiotics are given as soon as possible, if permanent damage to the lining of the joints is to be prevented. Once infection enters and damages the joint cartilage, permanent arthritis may follow. In some cases it may be helpful to drain the infected joint, and to flush it with fluids to remove as much of the infection and debris as possible, before injecting antibiotics directly into it. Joint-ill is always serious. If a foal becomes lame, the vet should always be called to examine it, just to rule out this possibility. Likewise, it is sensible to take the temperature of any foal that refuses to suck, or that seems 'sleepy', and to seek veterinary advice if it is higher than 38.9°C (102°F).

? MY HORSE IS NOT MOVING WELL: COULD THE TROUBLE BE IN ITS BACK?

Back problems are quite common in older horses, especially in those required to jump. Affected animals may show changes in their action rather than specific lameness. Their strides may be shortened, and they may fail to use their hind limbs properly. When moving in a circle, they can also be reluctant to bend the back. Back problems are frequently a cause of poor performance. Pinpointing the exact site of injury in the back is usually difficult, however — and often impossible. Horses have very strong and well-developed muscles running above and below the vertebrae. These can be strained when jumping. Horses also have high spines (vertebral processes) sticking out from the backbone — these can be felt along the horse's back. Thickening of the bone at the tip of these spines may mean that two neighbouring spines can impinge on one another, causing pinching and pain when the animal flexes its back. Because of the depth of the horse's body and the thickness of its back muscles, it is impossible to take very

clear X-ray pictures of the backbone, even when the horse is anaesthetised; although major changes — such as fractures or problems in the vertebral processes — can be seen, small changes in the vertebrae themselves cannot. As horses get older, the bones of the spine — especially in the lumbar region — tend to fuse together. Some discomfort may occur while new bone is being formed, but this ceases once the bones are joined. Injury to superficial back muscles can sometimes be detected using Faradic stimulation.

A site of frequent injury is at the junction of the backbone and pelvis. Such sacro-iliac damage is more common in large-framed horses with weak hindquarters, and causes chronic problems, including loss of performance and altered hind leg action (lack of impulsion from one, other, or both hind limbs). This joint may also be affected by trauma (falling), and the pelvis may appear unlevel when viewed from behind (the point of the hip and the highest point of the croup — tuber sacrale — may appear higher on one side than the other). This latter injury frequently settles down, and horses with unlevel quarters can perform perfectly adequately.

Horses with back problems are frequently reluctant to bend the spine. This can be detected by running a finger or a blunt object (like a spoon) along the animal's back, from the withers to the tail. A normal horse dips its back when pressure is applied along the middle of the back, and flexes it upwards when the same pressure is applied along the croup towards the tail. Similarly, when pressure is applied in a semicircular motion to the upper flank, below the backbone, a normal horse bends its back towards that side. A horse with a back injury may be reluctant to perform any of these movements, or may show signs of pain. A 'cold' back (dipping the back when the saddle is put on) is not usually a sign of back injury.

Treatment of back injuries usually involves rest, sometimes combined with anti-inflammatory drugs given by mouth. Specific treatments include surgical removal of the dorsal processes of vertebrae to relieve impinging spines. When a chronic sacro-iliac problem is diagnosed, rest is of little benefit. A gradually increasing exercise schedule is much more helpful, by strengthening and building up hindquarter muscles to 'support' this joint. A wide variety of alternative treatments is used for back injuries, with varying degrees of success.

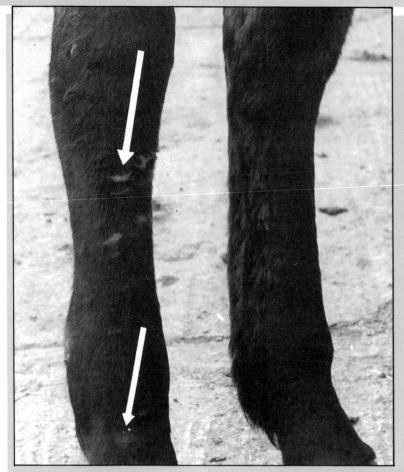

Lymphangitis (arrowed) is shown here by ulceration of the lymph glands *(above)*. In some instances, the leg will swell rather than ulcerate, and is clearly very tender. This condition most frequently affects the hind legs.

degree of damage. In cases of poor performance, similar blood tests aid the detection of very small degrees of muscle damage.

The exact biochemistry of this condition is not fully understood, but seems to be related to electrolyte imbalances. It may well be that different disturbances in the electrolyte balance within the body may produce similar clinical signs. Acute forms of this condition can be extremely painful, causing the animal great distress. It is important not to move a horse with azoturia if at all possible. Treatment involves the use of painkillers and intravenous electrolytes (in severe cases), and rest. Prevention is better achieved by paying close attention to feeding in relation to exercise — if a horse cannot be ridden for some reason, cut back its concentrate ration. In animals that are prone to this problem, adding electrolytes to the diet and ensuring they have an adequate salt intake may help prevent it from recurring.

? MY HORSE IS LAME, AND ONE HIND LEG IS VERY SWOLLEN: WHAT COULD THE PROBLEM BE?

Sporadic lymphangitis is a non-infectious condition, usually affecting one hind leg. It arises when a horse is receiving a full working feed ration, but for one reason or another (perhaps bad weather) is not exercised. The swelling is due to an accumulation of lymph under the skin. The entire leg may become very swollen, and the swelling can spread up into the groin. Sometimes a small skin wound can be detected, but in most cases there is no evidence of injury. The condition is very painful, and the animal is extremely reluctant to move. It will also have a high temperature, and will refuse to eat. Treatment involves antibiotic and cortisone injections, accompanied by diuretics. Painkillers such as 'bute' are also given because the horse must be made to walk and help disperse the swelling. Hot fomentation, to massage the limb, and cotton wool and leg bandages to try and restrict the swelling, are also helpful. Gentle exercise is necessary to reduce the swelling, which may take a week or more to disperse. Quite often, some permanent thickening of the leg occurs, and there is a tendency for this problem to recur in the same leg. The condition can be prevented by strict attention to feeding in relation to exercise. If, for any reason, a fit horse cannot be ridden, its concentrate

? WHAT IS AZOTURIA, AND HOW CAN IT BE AVOIDED?

This condition is known by many different names including 'tying up', 'setfast', 'Monday-morning disease' and equine rhabdomyolysis. It is in effect a syndrome involving muscle damage, which can arise in a variety of clinical situations. Symptoms range from mild muscle stiffness to extensive muscle damage, inability to move, and, possibly, to death. The mildest forms occur in horses that are overexercised after a period of rest. This usually happens when the animals have been fed a full working ration during a rest period. More severe signs can occur, and soon after exercise has ended, the horse or pony becomes very stiff and reluctant to move. It may later pass dark, brown-coloured urine. In the past, this sometimes happened when draughthorses were rested over the weekend, yet still received full working rations of feed. When

work was resumed, signs of azoturia developed — hence the name 'Monday-morning disease'. The discoloration of urine is due to the presence of the nitrogenous breakdown products of damaged muscle — hence the term 'azot-uria'. Individual horses (particularly mares) seem especially prone to this condition, and require very careful management of both exercise and diet.

Muscle damage can also occur as a result of prolonged strenuous exercise in an unfit horse — for example a horse at grass by itself, which panics and gallops round to exhaustion. Muscle damage can also be associated with prolonged travel (many hours in a horse-box or aeroplane) and during very long surgical operations under general anaesthesia. The muscle damage can easily be detected by blood tests. Diagnosis is based on the clinical signs, but blood tests are essential to measure enzymes released by damaged muscle, and thus to assess the

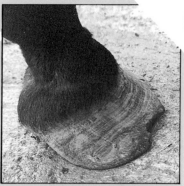

Regular examinations of a horse's shoes should be made. If they have been left on too long, they can cause problems. Often they begin to slip round, as seen above.

In the case shown on the left, **the clenches have risen** and are now standing proud of the hoof wall, and the lower part of the horn has begun to crack and break away. Even if the shoes are not worn down, the feet need regular attention from the blacksmith. The shoes should be taken off, and the feet trimmed and rasped before they are replaced.

ration should be cut back, and a laxative mash given instead. Puncture wounds on the lower limb may cause swelling of a hind leg, as can secondary infection of cracked heels. This does not usually spread above the hock. Slight filling of all four limbs may occur when horses (especially Thoroughbreds) are not exercised. Lymph fluid accumulates around the fetlock joints, and the horse is said to have 'humor'. There should be no confusion with lymphangitis, because several legs are involved, and there is no heat or lameness. The 'filling' of the legs disappears with exercise, or if the legs are bandaged when the horse is standing in the stable.

? MY HORSE IS LAME, AND THERE IS HEAT IN THE FOOT: WHAT IS THE MOST LIKELY CAUSE?

The most likely cause of heat in one foot, causing lameness, is pus in the foot. Infection enters through the sole and works its way through cracks in the horn (under-run sole) into the sensitive parts of the foot, where an abscess forms. Sometimes infection may enter as a result of a puncture wound (treading on a nail), or small pieces of stone may become wedged in the horn and work their way up the wall. Infection may also enter when a nail is driven too close to the sensitive layers of the foot during shoeing –

the horse is said to have 'nail bind' or to have been 'pricked'. The degree of lameness is variable.

Some idea of the position of the pus may be evident from the way the animal puts the affected foot to the ground. Thus, if pus is present under the heels, the horse may walk on the toe; or if pus is in the outside of the foot, it may walk with its leg out, to take the weight on the inner wall. It may be possible to judge where the pus is by pressing the sole with the thumb. Thumb pressure can also be helpful to detect puncture wounds in the sole and frog. Deep-seated infection does not respond to thumb pressure, but can usually be detected by a vet using special hoof-testing pincers.

Treatment involves making an opening in the sole to release the pus, using a hoof knife. The shoe is normally removed, and poulticing the foot for 24 hours beforehand greatly assists this process. Indeed, poulticing not only softens the horn and makes it easier to cut, but helps to bring a foot abscess to a head. An opening is made large enough to allow drainage, and the foot may be poulticed again, to help draw out the remainder of the infection. A glycerine and Epsom salts mixture is particularly good for this purpose. Lameness usually disappears soon after the poison has been released. A cotton wool plug is inserted into the hole to keep it clean until the hoof grows to fill it. If it

has been necessary to make a very large opening, a leather pad can be fitted, under the shoe, to keep earth out. Rarely, it may not be possible to find pus by searching the sole; the poison may have spread up the wall of the hoof. In this case repeated poulticing for several days is required to soften the hoof wall, and the poison eventually breaks out at the coronet. The process may take several weeks, but after poulticing for a few more days, it clears up, causing no further problems. Antibiotics are usually not much help for pus in the foot because they do not penetrate into the horn. However, infection may spread up the leg, and antibiotic treatment may be required in that case. Puncture wounds from very long nails in the frog region are serious; important structures (such as the deep digital flexor tendon) may be penetrated and become infected. A vet should always be consulted in this case, because antibiotic treatment may be necessary to prevent serious problems. If a horse is unvaccinated against tetanus, tetanus antitoxin must be given whenever a foot (or any other) puncture wound is discovered, and if it is necessary to cut out pus in the foot with a hoof knife. Other conditions can cause heat in horses' hoofs, but these affect more than one leg (laminitis), or produce only very slight heat (navicular disease).

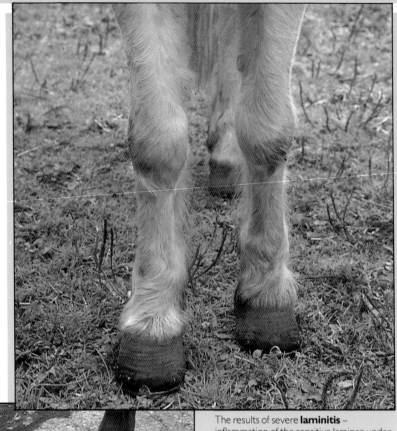

The results of severe **laminitis** – inflammation of the sensitive laminae under the wall of the foot *(above)*. Structural changes have occurred within the hoof, causing the sole to drop and the foot to look 'boxy'.

Close-up *(left)* of a hoof in a horse suffering from **laminitis**. The disease has resulted in pronounced rings round the horn.

unwelcome complication of treatment with drugs. Cortisone in horses sometimes has this undesirable effect.

In acute laminitis, signs include systemic distress and local changes in the foot. The horse invariably has a temperature, and is off its food. It stands with its forelegs well out in front of it, and the hind legs as far forward under the body as possible to take most of the body weight. There is heat in the feet, and a bounding pulse in the digital artery can usually be felt at the fetlock. It is often impossible to pick up a leg; even if it can be done, the slightest finger pressure on the sole is resented. In very acute cases, signs of shock quickly follow, and the horse may die within 12-24 hours. Another common complication in acute laminitis is separation of the sensitive laminae of the hoof wall from the pedal bone. This permits rotation of the pedal bone within the hoof, caused by the pull of the deep flexor tendon (which inserts on the pedal bone). When the pedal bone rotates, it presses down on the sole, which loses its normal concave shape as a result. The bone may penetrate the sole and become exposed to the air, with serious consequences.

Chronic laminitis is one of the most common causes of lameness in ponies. It occurs in successive years, when ponies (frequently overweight) are turned out in lush pasture during May and June. The signs are usually confined to the feet, and include the heat, pain and abnormal gait seen in acute laminitis. There is often alteration in the growth of horn, and rings of horn ('laminitic rings') are formed around the hoof wall. The toes become long, through lack of wear, and there is frequently a dropped sole, due to rotation of the pedal bone.

HOW CAN IT BE TREATED?

Treatment of acute laminitis requires measures to counteract shock, including intravenous fluids. Laxatives are usually given to clear out the bowels, and diuretics and cortisones can help reduce fluid retention. Recently a drug (Isoxsuprine) has become available, which specifically increases the blood flow to the feet and appears to be helpful in acute laminitis. Painkillers are also particularly important, to enable the horse to walk, to improve the circulation in its feet. Forced exercise may be required, and the further the horse goes, the better it usually moves. Rarely, it may be necessary to use nerve

WHAT CAUSES LAMINITIS?

Horses with laminitis usually have heat in all four feet, but particularly in the fore feet. They are reluctant to move, and may lie down, or walk on their heels to try to avoid putting pressure on the soles of the feet. The condition is due to inflammatory changes in the sensitive laminae of the hoofs, just below the hoof wall. This arises because of a congestion of blood within capillary vessels supplying the hoof. Research has shown that in laminitis, alterations in the blood-flow to the feet occurs. The capillary vessels below the coronet are bypassed, and blood accumulates there, damaging the sensitive laminae.

There are several different possible causes of laminitis. Some of these are associated with changes in the digestive system,

including over-eating on grain or concentrates — particularly horses that break into the feed store! It is thought that a chemical histamine, produced during the digestion of protein, is one substance that can be responsible for development of laminitis. Histamine is normally metabolized in the liver, but in over-eating the liver is unable to cope — histamine enters the circulation and affects the blood vessels in the feet. Lush grass is often associated with laminitis in ponies. Toxins from infections elsewhere in the body (such as pneumonia) may also produce a secondary laminitis. This sometimes happens shortly after foaling in mares that have retained a portion of the foetal membranes in the uterus, or have a uterine infection for some other reason. In this case, the laminitis is acute and often serious. Laminitis may also arise as an

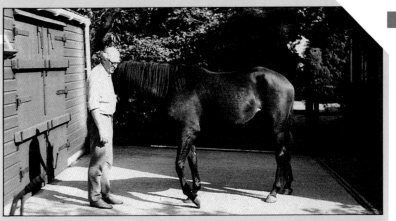

Resting a foreleg is nearly always the sign of trouble *(above)*. It is the characteristic stance of a horse with **navicular disease** – a corrosive ulcer of the navicular bone .

? WHAT ARE THE SIGNS OF NAVICULAR DISEASE?

Navicular disease is a form of chronic foreleg lameness, usually affecting both limbs. The onset is generally very gradual, and the lameness may be transient at first – the animal pulling out slightly lame, but becoming 'sound' during exercise. Often, the problem is only appreciated in one leg at first, and it is not until a nerve block is performed on that leg, that trouble in the opposite limb is noticed. Rarely, navicular disease may produce acute lameness in one leg. The disease usually develops in middle-aged horses (5-8 years) but can be seen in animals as young as yearlings. Larger breeds of horse are affected – the condition is practically unknown in animals under 14.2 hands.

The navicular bone is held tightly in place by ligaments, and acts as a fulcrum over which the deep digital flexor tendon (DDFT) runs. In severe cases of navicular disease, extensive changes occur on the back surface of the navicular bone (which is in contact with the DDFT). The cartilage is eroded, and a large area of degeneration may be present in the centre of the bone. There may also be changes in the DDFT associated with this, including thickening of the tendon and damage to the tendon sheath. In some serious cases, the DDFT may actually stick to the navicular bone. Also, spurs of new bone are often formed on the side edges of the navicular bone. All these changes can be seen on an X-ray of the foot, and such horses are permanently lame. In earlier or milder forms of the disease, lameness may be only slight. The horse tends to try to put more weight on the toe to avoid pressure on the navicular bone (which lies across the foot, just above the point of the frog). This results in a shortening of the stride. The foot is not put down flat – the toe landing slightly before the heels – which causes extra wear on the front of shoes and may cause stumbling on uneven ground. Less often the horse may rest in the box, 'pointing' one or other toe.

Recent research has shown that in navicular disease there is a decreased blood-flow within the navicular bone. This is thought to be responsible for much of the pain in the disease – tissue deprived of blood is always painful. As a result of this lack of blood, degeneration of bone within the navicular occurs, causing thinning of the bone, which can be seen on X-ray examination. At the same time, the body's attempt to heal this damage is to increase and enlarge blood vessels entering the base of the bone. Thus, in navicular disease, changes in the outline of blood channels at the base of the navicular bone can be seen on radiographic examination. X-rays of the foot can, therefore, be helpful in confirming a diagnosis in the early stages of the disease. This matter is not entirely straightforward, however, because some horses that are quite lame have little or no changes on X-ray, whereas other horses that have never been lame may show quite marked changes in the navicular bone. The probable explanation for this is that, because the pain and lameness in the early stages may be associated with a reduced blood supply, whether or not lameness is present depends on whether bypass blood vessels have nevertheless managed to supply the areas of bone of which the normal blood supply has been blocked.

? CAN IT BE TREATED?

Until recently, navicular disease was incurable although in some horses the condition could be temporarily relieved with painkillers such as phenylbutazone ('bute'). In many cases, the horse's working life ended, and it was frequently necessary to destroy it on humanitarian grounds because of pain from permanent lameness. The knowledge that a blockage of the blood vessels within the navicular bone was an important factor in the development of the disease has led to treatments which, for the first time, have been able to successfully treat this condition. The anticoagulant Warfarin (a common constituent of rat poisons) was the first drug to be used. This is given in the feed for several weeks, to increase the time it takes for blood to clot, thereby helping to reduce clots in the navicular bone. This drug is inexpensive but requires repeated blood sampling to check the blood-clotting time; an overdose could lead to fatal haemorrhages. The symptoms are successfully relieved in about 80% of cases, although some relapses occur when treatment ceases. Another drug – Isoxsuprine – is now also used for this disease: it has been found to have a specific action in increasing the blood flow in the digital arteries of horses. The drug is relatively more expensive, but does not have the side effects of Warfarin. Its success rate seems to be similar to Warfarin, although again relapses can occur when treatment

Surgical shoeing has also been used for many years for the treatment of this condition. In Scandinavia, considerable success has been achieved using an egg-bar shoe. This has a bar across the heel of the shoe, in the centre of which is an egg-shaped metal plate which presses up under the centre of the frog.

The all too common fault of long toes and low heels is particularly likely to predispose an animal to this problem because extra leverage from the DDFT puts more pressure on the navicular bone. It may be possible to prevent this disease by correcting this foot defect.

WHAT ARE SIDEBONES?

A horse's foot has two cartilages on either side, joined to the wings of the pedal bone. These cartilages move as the horse walks, and the tops of the cartilages can be felt just above the coronet (towards the heels) on each side of the leg. In older horses, they frequently harden and become bone (ossify); this is a normal ageing process. The hardened cartilages can be felt to have lost their 'spring', and this is what is called sidebones by horsemen. The condition is more common in draughthorses, rare in riding horses, and seldom causes lameness, unless the tip of the ossified cartilage is damaged (fractured by being trodden on). In younger horses, repeated concussion can cause pain, heat and soreness in tissues around the cartilages, and premature ossification can occur. Although sidebones are often seen on foot X-rays, lameness is rarely caused by them, and they are often blamed for trouble elsewhere. If pain and heat can be found at the correct spot on the coronet, the horse must be rested. Grooving the heels, done by a blacksmith, may permit more expansion of the heels. In the past, in cab horses, wounds caused by treading on the coronet sometimes resulted in infection of the lateral cartilages, a condition known as 'quittor' that was hard to treat; the infected cartilage discharged pus for many weeks.

WHAT IS THRUSH, AND HOW CAN I PREVENT IT?

Thrush is a degeneration of the outer layers of the frog in horses' feet. This becomes soft and emits a characteristic foetid smell. The condition arises from poor stable hygiene — failure to pick out the feet regularly. Moisture from impacted droppings softens the frog, and bacteria from faecal decomposition enter the softened frog tissue and begin to break it down. Thrush can be more of a problem in horses with contracted heels and a small frog because it is then easier for droppings to 'ball up' in the feet.

Treatment involves thorough cleaning of the frog and frog clefts, and the removal of all abnormal frog tissue (with a hoof knife). Antibiotic sprays, or iodoform liquid, can be used to dry the frog and clear up the infection. Alternatively, astringent powders (such as copper sulphate) can be packed around the frog, packed into its clefts, and held in place with cotton wool and a bandage. Rarely, infection may enter deeply into the frog and affect sensitive layers beneath, causing lameness. In this case, antibiotic treatment may be necessary. In past centuries, a similar foul-smelling decomposition of the sole occurred in conditions of very bad stable hygiene. This was known as 'canker'. A somewhat milder condition is encountered when hoof pads are left on too long, under shoes. The treatment is the same as for thrush.

WHAT IS PEDAL OSTEITIS?

Some horses that are particularly sensitive to uneven going ('sore-footed'), or that are slightly lame, are found on X-ray examination of the foot, to have less bone density than normal in the pedal bone. The edges of the bone appear roughened, due to new bone formation. This is usually more apparent towards the heels. Affected horses can show mild, chronic lameness in one, other, or both fore feet. The disease is thought to be the result of repeated concussion, following which new bone is formed as a result of inflammation of the pedal bone (osteitis). This presses on the sensitive layers of the sole. Larger horses with flat feet (that is, soles that are not concave) are much more likely to develop this trouble. Diagnosis is not always straightforward because horses with similar changes on X-ray, and showing no signs of lameness, are frequently encountered. It may be necessary to nerve-block the foot to confirm the diagnosis. Treatment involves giving the bone inflammation time to settle down, and reducing concussion. A spell at grass is often the best answer, and rubber or other synthetic anti-concussion pads fitted under the shoes can be very helpful.

MY HORSE HAS A SPLIT IN THE HOOF WALL: WHAT CAN I DO ABOUT IT?

A vertical split in the hoof wall is known as a 'sand crack' when it occurs at the front of the hoof, and a 'quarter crack' when it occurs at the heels. These can cause lameness either by pinching sensitive layers underneath as the horse moves, or less often, by allowing infection to enter. New horn is formed at the coronet, growing down the wall at approximately one centimetre (half an inch) a month. Cracks thus 'grow out' if the split can be prevented from spreading up the wall into the horn above. This is done by rasping a groove across the wall at the top of the crack to stop it spreading upwards, at the same time trying to minimize movement of the horn on either side of the split. It is essential that a shoe is fitted; these frequently have two clips (similar to toe clips) made on either side of the split to hold the horn firmly in place. At the time of shoeing, a large V-shaped cut is made in the sole at the base of the split. This is often sufficient to restrict movement and prevent the split from spreading. With larger, deeper cracks, and especially with quarter cracks, greater immobilization may be required. Some blacksmiths fit staples across the crack to do this, but more often, acrylic resins are used to fill in the gap and immobilize it until it can grow out. Wounds at the coronet are often serious. If the coronary band is damaged, normal horn may not be produced afterwards at the site. This may lead to a permanent growth of a cracked hoof wall which will require a very skilled farrier to keep the horse 'sound'.

The sole of an unshod foot, showing a prominent, healthy frog *(left)*.

A foot with several vertical cracks – known as **sandcracks** – in the hoof wall, spreading upwards *(below)*. An 'X' has been rasped at the top of a toe crack, and a horizontal line at the top of a quarter crack, to try to prevent them spreading: They should then grow out.

LONGEVITY

How long a horse lives will depend greatly on the kind of life it has led. As it gets older it will begin to slow down and will need regular care and attention to ensure it is not suffering in any way.

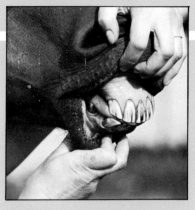

A clear indication of **old age**. This is the mouth of a 27-year-old horse *(above)*, in which the pale gums indicate anaemia and the incisor teeth are well worn. Dental problems are a common cause of lost body condition in old age.

 HOW LONG DO HORSES LIVE?

The oldest recorded age for a horse is 62 years, for a Cleveland Bay cross gelding called Old Billy, which was foaled in 1760. Horses and ponies usually live for 20 to 25 years, and those that live on up to 30 years are not uncommon. However, very few live much beyond this age. In terms of life-span, one year of a horse's life is equivalent to approximately three years of a human life.

WHAT PROBLEMS DO HORSES ENCOUNTER IN OLD AGE?

The main problems in old horses are arthritis, muscle stiffness, and difficulty in maintaining body condition, all of which are more of a problem in the winter when it is necessary to keep elderly horses stabled, at least at night. Dental problems are also more common in old age; teeth should be checked at least twice a year. Old horses rarely suffer from heart conditions or cancer.

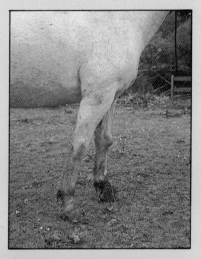

Arthritis – a common problem in old age. In this case *(above)* the horse has pronounced swelling and pain in the left knee.

HOW DO I KNOW WHEN IT IS TIME TO HAVE MY OLD HORSE PUT DOWN?

It is often very difficult for an owner to decide when is the best time to put down a horse that has been a good and faithful servant. Unfortunately, old horses seldom die suddenly. More often they gradually decline, deteriorating in health and body condition over a lengthy period. Winter is the hardest time for them, and each autumn a decision should be made whether or not it is kind to keep a retired horse through the winter. Because of the difficulty of making a detached judgement, it is very helpful to ask someone else's opinion on this subject. It is often easier for someone else, who does not see the horse every day, to gauge the animal's deterioration and make an objective assessment of its general health and well-being.

Retirement should, wherever possible, be active. A horse that is no longer able to compete can still for many years enjoy being ridden out for an occasional walk, and a change of scene – even if only on a leading rein – benefits all old horses.

There is little kindness in keeping an old horse or pony by itself, in a field, with mud up to its hocks, just for the sake of keeping it alive. It is preferable to have a horse destroyed while it is still in good condition and 'enjoying' life, rather than waiting until it becomes very thin and weak or in permanent pain from arthritis.

WHY ARE HORSES USUALLY DESTROYED BY SHOOTING THEM?

The problem with euthanasia in horses is disposal of the body afterwards. If a horse is given an overdose of barbiturates by injection (in a manner similar to putting a dog to sleep) to destroy it, the carcass will have to be buried; it cannot be disposed of by a knacker. Most owners do not have the space or the facilities to bury a horse, and for this reason, horses being put down are normally shot, and disposed of by the knacker.

The thought of this is not pleasant for any owner – but the horse will be unperturbed, quite unaware of what is to happen, and death is instantaneous.

Retirement. Happy and content this horse *(Left)* may be in the summer, but will the same be true in the winter?

GLOSSARY

Azoturia Condition affecting the muscular system, also known as Monday-morning disease.

Aged Description of a horse that is seven years old or more.

BHS British Horse Society.

Blanket clip Removal of the coat from the neck, belly and tops of legs, leaving a blanket shape over the back.

Box-walking A stable vice, when a horse paces endlessly around its stable.

Brushing The striking of the fore or hind leg with the foot of the opposite leg.

Canker Infection of the sole of the foot.

Carpitis Sore shins or knee troubles in young Thoroughbreds, caused by overwork and strain.

Cast When a horse is lying down in the stable, unable to get up, because it has somehow become wedged against the wall.

Chaff Finely chopped hay.

Chestnut Small piece of horn on inside of horse's legs. Also a coat colour.

Clinch End of nail holding shoe in place, visible round the front of the hoof. Known also as clench.

Cold-blooded Any of the heavy horse breeds, or those tending towards heaviness.

Colic Disorder in the horse's digestive organs, causing acute abdominal pain.

Colostrum The first milk, containing vital antibodies, that a foal drinks from its dam.

Colt Male foal.

Condylarthra Primitive mammal that lived 75 million years ago: the horse's earliest ancestor.

COPD Chronic Obstructive Pulmonary Disease, known previously as 'broken wind'.

Covering A stud term — a mare is covered when she is mated.

Crib-biting One of the stable vices — when a horse continually catches hold of an object and sucks air into its stomach.

Cross-breed The offspring of the mating of two different breeds of horse.

Cubes Complete concentrated horse and pony feed, known also as nuts.

Dam Mother horse.

Deep litter System of bedding, in which only surface soiled bedding is removed each day, and more clean bedding is spread on top.

Dish-face Convex profile to a horse's face.

Dishing Faulty action, in which the front feet move in an almost circular motion.

Double bridle Bridle comprising a snaffle (bridoon) and curb bit.

Dressing (the foot) Trimming of the foot, done every four to six weeks by the farrier.

Dressing (the navel) Application of antibiotic powder to the ruptured umbilical cord of a newly born foal.

Equus caballus Generic term for all horses.

Equine rhinopneumonitis Equine Herpes Virus Type I, the most common virus to give 'flu-like symptoms.

Ergot Small horny growth on the skin of the back of the fetlock joints.

FEI Fédération Equestre Internationale.

Filly Female horse or pony under four years old.

First-stage labour Opening of cervix and emergence of amniotic sac during early labour.

Foal Male or female horse or pony, up to one year old.

Forging Striking of the sole of the forefoot with the toe of the hind foot.

Frog V-shaped part of sole of foot which acts as a shock absorber during motion.

Galvayne's groove A dark brown mark which appears on the corner incisor teeth when a horse is ten years old. It grows down the tooth as the horse gets older.

Gamgee Gauze-covered cotton wool, used under stable and travelling bandages for warmth and protection.

Gastrophilus intestinalis Adult botflies — a horse parasite.

Gelding A castrated male horse or pony.

Girth galls Sores on the underbelly or behind the elbows, caused by the girth rubbing.

Hackamore A bitless bridle.

Hand Unit of measure equalling 10cm (4in), used to estimate a horse or pony's height.

Hard feed Any type of corn or concentrated feed.

Horse Any breed of *Equus caballus* measuring over 14.2 hh (hands high). See also **Pony.**

Hunter clip Removal of all the hair from a horse's body except for a saddle mark and that on the legs.

Joint Measuring Scheme A recognized scheme for measuring horses and ponies, approved and sponsored by various equestrian societies and associations.

Loose box Self-contained indoor accommodation for horses and ponies.

Lungeing The training of a young horse on a circle from the ground, using a long lunge rein.

Lungworm A species of horse parasite that causes respiratory problems unless preventive treatment is given.

Lymphangitis A circulatory disorder —

inflammation of the lymph vessels.

Mare Female horse or pony.

Martingale Item of tack used to increase control of a horse's head, or to alter the pull of the reins.

Nappy Term used to describe disobedience — shying, tossing the head etc — in a horse.

Native ponies Any of the pony breeds native to the British Isles — Exmoor, Dartmoor, Welsh Mountain, New Forest, Fell, Dale, Connemara or Shetland.

Nearside The left side of a horse or pony.

New Zealand rug Heavy waterproof rug with woollen lining, for horses or ponies out at grass in the winter.

Numnah A pad made of felt, sheepskin or soft fibre to fit under the saddle to prevent undue pressure.

Oestrus Breeding period of a mare.

Offside Right side of a horse or pony.

Over-reaching Faulty action, in which the heel of the front feet is struck by the toe of the hind.

Plaiting Faulty action, in which the foot moves across and inwards to land more or less in front of the opposite foot. Also **plaiting** of the mane and tail for showing.

Pleohippus The first single-toed ancestor of the horse which lived 10 million years ago.

Points Term used to describe the different parts of a horse's anatomy. Also used in connection with colour to describe the mane, tail and lower legs.

Pony Any breed of *Equus caballus* measuring less than 14.2 hh.

Pulling (mane and tail) The thinning of the hairs of the mane and top of the tail by carefully and selectively pulling them out.

Quartering Brief grooming, given before a horse is taken out on exercise.

Rasping (of feet) Filing of the outer hoof, done by the blacksmith as he trims the feet and reshoes a horse.

Rasping (of teeth) Filing down teeth to remove sharp points that are causing pain, resulting in feeding and digestive disorders.

Ringworm Contagious skin disease, caused by a fungus. Hair loss occurs in little round patches.

Roman nose Convex profile to the face.

Saddle galls Sores on the back under the saddle, caused by an ill-fitting saddle, or one that rubs a sensitive back.

Salt-lick Shaped block of salt, fitted into a holder for a horse in the stable or paddock.

Scour, to Suffer from diarrhoea, particularly foals.

Second-stage labour Rupture of the amniotic sac, followed by period of straining, ending with the birth of the foal.

Setting fair (of coat) Superficial grooming to smooth the coat after removing rugs.

Setting fair (of bed) Putting down a horse's bed before bedding down at night.

Single bridle Bridle with just one bit — most frequently a snaffle.

Sire Father horse.

Stable vices Various bad habits acquired by horses, usually through boredom at being stabled.

Stale, to To urinate.

Stallion Uncastrated male horse.

Stalls Indoor equine accommodation, where horses are kept in three-sided compartments, separated from one another by side walls.

Strapping Thorough grooming of a stabled horse after exercise.

Strongylus vulgaris Migrating larvae of the horse parasite, the red worm.

Sweet itch A condition of the mane and tail, causing intense irritation.

Tack Items of saddlery and equipment used in the riding and training of horses.

Three-quarter bred Offspring of a Thoroughbred mare or stallion and a Thoroughbred-cross mare or stallion.

Thrush A degenerative condition of the outer layers of the frog in a horse's feet.

Trace-clip The removal of hair in a line down the underside of the neck, the belly and the top of the legs.

Twitch A stick with soft rope attached, placed over the upper lip and twisted until the lip is firmly held. It acts as a restraint to a horse and is used for specific purposes, such as during clipping.

Unthriftiness Failure to thrive in normal conditions.

Vetting Examination given to a horse by a vet to determine soundness etc.

Warm-blooded Horses possessing Arabian or Thoroughbred blood in their ancestry.

Weaving One of the stable vices — the horse rocks to and fro continually, shifting his weight from leg to leg.

Wind-sucking One of the stable vices — the horse sucks air into its lungs.

Wolf teeth First premolar teeth, occurring usually in the upper jaw. They are often very sharp and can cause considerable pain.

Yearling Male or female horse, between one and two years old.

INDEX

ACKNOWLEDGEMENTS

The majority of the photographs for this book were taken by
Bob Langrish. Many thanks to Ray Saunders for providing the
pictures on pages 104, 105, 123 *(bottom right)*, 134, 135, 137
(bottom), 143, 147, 148 *(right)*, 149, 150, 151, 153. Further illustrations are courtesy of Hazel Edington.